THE NATURE OF MATHEMATICS

THE NATURE OF MATHEMATICS

Philip E. B. Jourdain

DOVER PUBLICATIONS, INC.
Mineola, New York

Bibliographical Note

This Dover edition, first published in 2007, is an unabridged republication of the work included in Volume I (pages 2-71) of *The World of Mathematics*, edited by James R. Newman, which was published in 2000 by Dover Publications, Inc. (ISBN: 0-486-41153-2). Jourdain's book was originally published by T. C. and E. C. Jack, London, and Dodge Publishing Company, New York, in 1913. The Preface to the current edition, also from *The World of Mathematics*, was written by James R. Newman.

International Standard Book Number: 0-486-45885-7

Manufactured in the United States of America
Dover Publications, Inc., 31 East 2nd Street, Mineola, N.Y. 11501

PREFACE

PHILIP E. B. JOURDAIN (1879–1919), whose little book on the nature of mathematics is here reproduced in its entirety, was a logician, a philosopher and a historian of mathematics. To each of these subjects he brought a fresh outlook and a remarkably penetrating and creative intelligence. He was not yet forty when he died and from adolescence had been afflicted by a terrible paralytic ailment (Friedrick's ataxia) which gradually tightened its grip upon him. Yet he left behind a body of work that influenced the development of both mathematical logic and the history of science.

Jourdain, the son of a Derbyshire vicar, was educated at Cheltenham College and at Cambridge. The few years during which he was able to enjoy the normal pleasures of boyhood—long walks were his special delight—are described in a poignant memoir by his younger sister, Millicent, who suffered from the same hereditary disease. In 1900 the brother and sister went to Heidelberg to seek medical help. While at the hospital he began in earnest his study of the history of mathematics. "We had," wrote Millicent, "what was to be nearly our last bit of walking together here." The treatment was unavailing and when they returned to England, Jourdain could no longer walk or stand or even hold a pencil without difficulty. Nevertheless, he undertook with great energy and enthusiasm the first of a series of mathematical papers which established his reputation. Among his earlier writings were studies of Lagrange's use of differential equations, the work of Cauchy and Gauss in function theory, and conceptual problems of mathematical physics.[1] Between 1906 and 1912 he contributed to the *Archiv der Mathematik und Physik* a masterly group of papers on the mathematical theory of transfinite numbers, a subject in which he was always deeply interested. In the same period the *Quarterly Journal of Mathematics* published a group of essays on the development of the theories of mathematical logic and the principles of mathematics. Jourdain was an editor of *Isis* and the *Monist*, in whose pages appeared his articles on Leibniz, Napier, Hooke, Newton, Galileo, Poincaré and Dedekind. He edited reprints of works by De Morgan, Boole, Georg Cantor, Lagrange, Jacobi, Gauss and Ernst Mach; he wrote a brilliant and witty book, *The Philosophy of Mr. B*rtr*nd R*ss*ll*, dealing with Russell's analysis of the problems of logic and the foundations of mathematics; he took out a patent covering an invention of a "silent engine" (I have been unable to

[1] Bibliographies of Jourdain's writings appear in *Isis*, Vol. 5, 1923, pp. 134–136, and in the *Monist*, Vol. 30, 1920, pp. 161–182.

discover what this machine was) and he wrote poems and short stories which never got published. In 1914, at the height of his powers, he was producing enough "to keep two typists busy all day."

The distinctive qualities of Jourdain's thought were its independence and its cutting edge. He was renowned for his broad scholarship in the history and philosophy of science, but he was more than a scholar. Never content with comprehending all that others had said about a problem, he had to work it through in his own way and overcome its difficulties by his own methods. This led him to conclusions peculiarly his own. They are not always satisfactory but they always deserve close attention: Jourdain rarely failed to uncover points overlooked by less subtle and original investigators.

The Nature of Mathematics reflects his excellent grasp of the subject, his at times oblique but always rewarding approach to logic and mathematics, his wit and clear expression. He had sharpened his thinking on some of the hardest and most baffling questions of philosophy and had achieved an orderly understanding of them which he was fully capable of imparting to the attentive reader. The book is not a textbook collection of methods and examples, but an explanation of "how and why these methods grew up." It discusses concepts which are widely used even in elementary arithmetic, geometry and algebra—negative numbers, for example—but far from widely comprehended. It presents also a careful treatment of "the development of analytical methods and certain examinations of principles." There are at least two other excellent popularizations of mathematics, A. N. Whitehead's celebrated *Introduction to Mathematics* [2] and the more recent *Mathematics for the General Reader* by E. C. Titchmarsh.[3] Both books can be recommended strongly, the first as a characteristic, immensely readable work by one of the greatest of twentieth-century philosophers; the second as a first-class mathematician's lucid, unhurried account of the science of numbers from arithmetic through the calculus. Jourdain's book follows a somewhat different path of instruction in that it emphasizes the relation between mathematics and logic. It is the peer of the other two studies and has for the anthologist the additional appeal of being unjustly neglected and out of print. "I hope that I shall succeed," says Jourdain in his introduction, "in showing that the process of mathematical discovery is a living and a growing thing." In this attempt he did not fail.

[2] Oxford University Press, New York, 1948.
[3] Hutchinson's University Library, London, n. d.

Pure mathematics consists entirely of such asseverations as that, if such and such a proposition is true of anything, *then such and such another proposition is true of that thing. It is essential not to discuss whether the first proposition is really true, and not to mention what the anything is of which it is supposed to be true. . . . If our hypothesis is about* anything *and not about some one or more particular things, then our deductions constitute mathematics. Thus mathematics may be defined as the subject in which we never know what we are talking about, nor whether what we are saying is true.*

—BERTRAND RUSSELL

The Nature of Mathematics

By PHILIP E. B. JOURDAIN

CONTENTS

INTRODUCTION

AN eminent mathematician once remarked that he was never satisfied with his knowledge of a mathematical theory until he could explain it to the next man he met in the street. That is hardly exaggerated; however, we must remember that a satisfactory explanation entails duties on both sides. Any one of us has the right to ask of a mathematician, "What is the use of mathematics?" Any one may, I think and will try to show, rightly suppose that a satisfactory answer, if such an answer is anyhow possible, can be given in quite simple terms. Even men of a most abstract science, such as mathematics or philosophy, are chiefly adapted for the ends of ordinary life; when they think, they think, at the bottom, like other men. They are often more highly trained, and have a technical facility for thinking that comes partly from practice and partly from the use of the contrivances for correct and rapid thought given by the signs and rules for dealing with them that mathematics and modern logic provide. But there is no real reason why, with patience, an ordinary person should not understand, speaking broadly, what mathematicians do, why they do it, and what, so far as we know at present, mathematics is.

Patience, then, is what may rightly be demanded of the inquirer. And this really implies that the question is not merely a rhetorical one—an expression of irritation or scepticism put in the form of a question for the sake of some fancied effect. If Mr. A. dislikes the higher mathematics because he rightly perceives that they will not help him in the grocery business, he asks disgustedly, "What's the use of mathematics?" and does not wait for an answer, but turns his attention to grumbling at the lateness of his dinner. Now, we will admit at once that higher mathematics is of no more use in the grocery trade than the grocery trade is in the navigation of a ship; but that is no reason why we should condemn mathematics as entirely useless. I remember reading a speech made by an eminent surgeon, who wished, laudably enough, to spread the cause of elementary surgical instruction. "The higher mathematics," said he with great satisfaction to himself, "do not help you to bind up a broken leg!" Obviously they do not; but it is equally obvious that surgery does not help us to add up accounts; . . . or even to think logically, or to accomplish the closely allied feat of seeing a joke.

To the question about the use of mathematics we may reply by pointing out two obvious consequences of one of the applications of mathematics: mathematics prevents much loss of life at sea, and increases the commercial prosperity of nations. Only a few men—a few intelligent philosophers and more amateur philosophers who are not highly intelligent—would doubt if these two things were indeed benefits. Still, probably, all of us act as if we thought that they were. Now, I do not mean that mathematicians go about with life-belts or serve behind counters; they do not usually do so. What I mean I will now try to explain.

Natural science is occupied very largely with the prevention of waste of the labour of thought and muscle when we want to call up, for some purpose or other, certain facts of experience. Facts are sometimes quite useful. For instance, it is useful for a sailor to know the positions of the stars and sun on the nights and days when he is out of sight of land. Otherwise, he cannot find his whereabouts. Now, some people connected with a national institution publish periodically a *Nautical Almanac* which contains the positions of stars and other celestial things you see through telescopes, for every day and night years and years ahead. This *Almanac*, then, obviously increases the possibilities of trade beyond coasting-trade, and makes travel by ship, when land cannot be sighted, much safer; and there would be no *Nautical Almanac* if it were not for the science of astronomy, and there would be no practicable science of astronomy if we could not organise the observations we make of sun and moon and stars, and put hundreds of observations in a convenient form and in a little space—in short, if we could not economise our mental or bodily activity by remembering or carrying about two or three little formulæ instead of fat books full of details; and, lastly, we could not economise this activity if it were not for mathematics.

Just as it is with astronomy, so it is with all other sciences—both those of Nature and mathematical science: the very essence of them is the prevention of waste of the energies of muscle and memory. There are plenty of things in the unknown parts of science to work our brains at, and we can only do so efficiently if we organise our thinking properly, and consequently do not waste our energies.

The purpose of this little volume is not to give—like a text-book—a collection of mathematical methods and examples, but to do, firstly, what text-books do not do: to show how and why these methods grew up. All these methods are simply means, contrived with the conscious or unconscious end of economy of thought-labour, for the convenient handling of long and complicated chains of reasoning. This reasoning, when applied to foretell natural events, on the basis of the applications of mathematics, as sketched in the fourth chapter, often gives striking results. But the methods of mathematics, though often suggested by natural events, are purely logical. Here the word "logical" means something more than the traditional doctrine consisting of a series of extracts from the science of reasoning, made by the genius of Aristotle and frozen into a hard body of doctrine by the lack of genius of his school. Modern logic is a science which has grown up with mathematics, and, after a period in which it moulded itself on the model of mathematics, has shown that not only the reasonings but also conceptions of mathematics are logical in their nature.

In this book I shall not pay very much attention to the details of the elementary arithmetic, geometry, and algebra of the many text-books, but shall be concerned with the discussion of those conceptions—such as that of negative number—which are used and not sufficiently discussed in these books. Then, too, I shall give a somewhat full account of the development of analytical methods and certain examinations of principles.

I hope that I shall succeed in showing that the process of mathematical discovery is a living and a growing thing. Some mathematicians have lived long lives full of calm and unwavering faith—for faith in mathematics, as

I will show, has always been needed—some have lived short lives full of burning zeal, and so on; and in the faith of mathematicians there has been much error.

Now we come to the second object of this book. In the historical part we shall see that the actual reasonings made by mathematicians in building up their methods have often not been in accordance with logical rules. How, then, can we say that the reasonings of mathematics are logical in their nature? The answer is that the one word "mathematics" is habitually used in two senses, and so, as explained in the last chapter, I have distinguished between "mathematics," the methods used to discover certain truths, and "Mathematics," the truths discovered. When we have passed through the stage of finding out, by external evidence or conjecture, how mathematics grew up with problems suggested by natural events, like the falling of a stone, and then how something very abstract and intangible but very real separated out of these problems, we can turn our attention to the problem of the nature of Mathematics without troubling ourselves any more as to how, historically, it gradually appeared to us quite clearly that there is such a thing at all as Mathematics—something which exists apart from its application to natural science. History has an immense value in being suggestive to the investigator, but it is, logically speaking, irrelevant. Suppose that you are a mathematician; what you eat will have an important influence on your discoveries, but you would at once see how absurd it would be to make, say, the momentous discovery that 2 added to 3 makes 5 depend on an orgy of mutton cutlets or bread and jam. The methods of work and daily life of mathematicians, the connecting threads of suggestion that run through their work, and the influence on their work of the allied work of others, all interest the investigator because these things give him examples of research and suggest new ideas to him; but these reasons are psychological and not logical.

But it is as true as it is natural that we should find that the best way to become acquainted with new ideas is to study the way in which knowledge about them grew up. This, then, is what we will do in the first place, and it is here that I must bring my own views forward. Briefly stated, they are these. Every great advance in mathematics with which we shall be concerned here has arisen out of the needs shown in natural science or out of the need felt to connect together, in one methodically arranged whole, analogous mathematical processes used to describe different natural phenomena. The application of logic to our system of descriptions, which we may make either from the motive of satisfying an intellectual need (often as strong, in its way, as hunger) or with the practical end in view of satisfying ourselves that there are no hidden sources of error that may ultimately lead us astray in calculating future or past natural events, leads at once to those modern refinements of method that are regarded with disfavour by the old-fashioned mathematicians.

In modern times appeared clearly—what had only been vaguely suspected before—the true nature of Mathematics. Of this I will try to give some account, and show that, since mathematics is logical and not psychological in its nature, all those petty questions—sometimes amusing and often tedious—of history, persons, and nations are irrelevant to Mathematics in itself. Mathematics has required centuries of excavation, and the process of excavation is not, of course, and never will be, complete. But we see enough now of what has been excavated clearly to distinguish between it and the tools which have been or are used for excavation. This confusion, it should be noticed, was never made by the excavators themselves, but only by some of the philosophical onlookers who reflected on what was being done. I hope and expect that our reflections will not lead to this confusion.

CHAPTER I

THE GROWTH OF MATHEMATICAL SCIENCE IN ANCIENT TIMES

IN the history of the human race, inventions like those of the wheel, the lever, and the wedge were made very early—judging from the pictures on ancient Egyptian and Assyrian monuments. These inventions were made on the basis of an instinctive and unreflecting knowledge of the processes of nature, and with the sole end of satisfaction of bodily needs. Primitive men had to build huts in order to protect themselves against the weather, and, for this purpose, had to lift and transport heavy weights, and so on. Later, by reflection on such inventions themselves, possibly for the purposes of instruction of the younger members of a tribe or the newly-joined members of a guild, these isolated inventions were classified according to some analogy. Thus we see the same elements occurring in the relation of a wheel to its axle and the relation of the arm of a lever to its fulcrum—the same weights at the same distance from the axle or fulcrum, as the case may be, exert the same power, and we can thus class both instruments together in virtue of an analogy. Here what we call "scientific" classification begins. We can well imagine that this pursuit of science is attractive in itself; besides helping us to communicate facts in a comprehensive, compact, and reasonably connected way, it arouses a purely intellectual interest. It would be foolish to deny the obvious importance to us of our bodily needs; but we must clearly realise two things:—(1) The intellectual need is very strong, and is as much a fact as hunger or thirst; sometimes it is even stronger than bodily needs—Newton, for instance, often forgot to take food when he was engaged with his discoveries. (2) Practical results of value often follow from the satisfaction of intellectual needs. It was the satisfaction of certain intellectual needs in the cases of Maxwell and Hertz that ultimately led to wireless telegraphy; it was the satisfaction of some of Faraday's intellectual needs that made the dynamo and the electric telegraph possible. But many of the results of strivings after intellectual satisfaction have as yet no obvious bearing on the satisfaction of our bodily needs. However, it is impossible to tell whether or no they will always be barren in this way. This gives us a new point of view from which to consider the question, "What is the use of mathematics?" To condemn branches of mathematics because their results cannot obviously be applied to some practical purpose is short-sighted.

The formation of science is peculiar to human beings among animals. The lower animals sometimes, but rarely, make isolated discoveries, but never seem to reflect on these inventions in themselves with a view to rational classification in the interest either of the intellect or of the indirect furtherance of practical ends. Perhaps the greatest difference between man and the lower animals is that men are capable of taking circuitous paths for the attainment of their ends, while the lower animals have their minds so filled up with their needs that they try to seize the object they want, or remove that which annoys them, in a direct way. Thus, monkeys often vainly snatch at things they want, while even savage men use catapults or snares or the consciously observed properties of flung stones.

The communication of knowledge is the first occasion that compels distinct reflection, as everybody can still observe in himself. Further, that which the old members of a guild mechanically pursue strikes a new member as strange, and thus an impulse is given to fresh reflection and investigation.

When we wish to bring to the knowledge of a person any phenomena or processes of nature, we have the choice of two methods: we may allow the person to observe matters for himself, when instruction comes to an end; or, we may describe to him the phenomena in some way, so as to save him the trouble of personally making anew each experiment. To describe an event—like the falling of a stone to the earth—in the most comprehensive and compact manner requires that we should discover what is constant and what is variable in the processes of nature; that we should discover the same law in the moulding of a tear and in the motions of the planets. This is the very essence of nearly all science, and we will return to this point later on.

We have thus some idea of what is known as "the economical function of science." This sounds as if science were governed by the same laws as the management of a business; and so, in a way, it is. But whereas the aims of a business are not, at least directly, concerned with the satisfaction of intellectual needs, science—including natural science, logic, and mathematics—uses business methods consciously for such ends. The methods are far wider in range, more reasonably thought out, and more intelligently applied than ordinary business methods, but the principle is the same. And this may strike some people as strange, but it is nevertheless true: there appears more and more as time goes on a great and compelling beauty in these business methods of science.

The economical function appears most plainly in very ancient and modern science. In the beginning, all economy had in immediate view the satisfaction simply of bodily wants. With the artisan, and still more so with the investigator, the most concise and simplest possible knowledge of a given province of natural phenomena—a knowledge that is attained with the least intellectual expenditure—naturally becomes in itself an aim; but though knowledge was at first a means to an end, yet, when the mental motives connected therewith are once developed and demand their satisfaction, all thought of its original purpose disappears. It is one great object of science to replace, or save the trouble of making, experiments, by the reproduction and anticipation of facts in thought. Memory is handier than experience, and often answers the same purpose. Science is communicated by instruction, in order that one man may profit by the experience of another and be spared the trouble of accumulating it for himself; and thus, to spare the efforts of posterity, the experiences of whole generations are stored up in libraries. And further, yet another function of this economy is the preparation for fresh investigation.[1]

The economical character of ancient Greek geometry is not so apparent as that of the modern algebraical sciences. We shall be able to appreciate this fact when we have gained some ideas on the historical development of ancient and modern mathematical studies.

The generally accepted account of the origin and early development of geometry is that the ancient Egyptians were obliged to invent it in order to restore the landmarks which had been destroyed by the periodical inundations of the Nile. These inundations swept away the landmarks in the valley of the river, and, by altering the course of the river, increased or decreased the taxable value of the adjoining lands, rendered a tolerably accurate system of surveying indispensable, and thus led to a systematic study of the subject by the priests. Proclus (412–485 A.D.), who wrote a summary of the early history of geometry, tells this story, which is also told by Herodotus, and observes that it is by no means strange that the invention of the sciences should have originated in practical needs, and that, further, the transition from perception with the senses to reflection,

[1] *Cf.* pp. 5, 13, 15, 16, 42, 71.

and from reflection to knowledge, is to be expected. Indeed, the very name "geometry"—which is derived from two Greek words meaning *measurement of the earth*—seems to indicate that geometry was not indigenous to Greece, and that it arose from the necessity of surveying. For the Greek geometricians, as we shall see, seem always to have dealt with geometry as an abstract science—to have considered *lines* and *circles* and *spheres* and so on, and not the rough pictures of these abstract ideas that we see in the world around us—and to have sought for propositions which should be absolutely true, and not mere approximations. The name does not therefore refer to this practice.

However, the history of mathematics cannot with certainty be traced back to any school or period before that of the Ionian Greeks. It seems that the Egyptians' geometrical knowledge was of a wholly practical nature. For example, the Egyptians were very particular about the exact orientation of their temples; and they had therefore to obtain with accuracy a north and south line, as also an east and west line. By observing the points on the horizon where a star rose and set, and taking a plane midway between them, they could obtain a north and south line. To get an east and west line, which had to be drawn at right angles to this, certain people were employed who used a rope ABCD, divided by knots or marks at B and C, so that the lengths AB, BC, CD were in the proportion 3 : 4 : 5. The length BC was placed along the north and south line, and pegs P and Q inserted at the knots B and C. The piece BA (keeping it stretched all the time) was then rotated round the peg P, and similarly the piece CD was rotated round the peg Q, until the ends A and D coincided; the point thus indicated was marked by a peg R. The result was to form a triangle PQR whose angle at P was a right angle, and the line PR would give an east and west line. A similar method is constantly used at the present time by practical engineers, and by gardeners in marking tennis courts, for measuring a right angle. This method seems also to have been known to the Chinese nearly three thousand years ago, but the Chinese made no serious attempt to classify or extend the few rules of arithmetic or geometry with which they were acquainted, or to explain the causes of the phenomena which they observed.

The geometrical theorem of which a particular case is involved in the method just described is well known to readers of the first book of Euclid's *Elements*. The Egyptians must probably have known that this theorem is true for a right-angled triangle when the sides containing the right angle are equal, for this is obvious if a floor be paved with tiles of that shape. But these facts cannot be said to show that geometry was then studied as a science. Our real knowledge of the nature of Egyptian geometry depends mainly on the Rhind papyrus.

The ancient Egyptian papyrus from the collection of Rhind, which was written by an Egyptian priest named Ahmes considerably more than a thousand years before Christ, and which is now in the British Museum, contains a fairly complete applied mathematics, in which the measurement of figures and solids plays the principal part; there are no theorems properly so called; everything is stated in the form of problems, not in general terms, but in distinct numbers. For example: to measure a rectangle the sides of which contain two and ten units of length; to find the surface of a circular area whose diameter is six units. We find also in it indications for the measurement of solids, particularly of pyramids, whole and truncated. The arithmetical problems dealt with in this papyrus—which, by the way, is headed "Directions for knowing all dark things"—contain some very interesting things. In modern language, we should say that the first part deals with the reduction of fractions whose numerators are 2 to a sum of fractions each of whose numerators is 1. Thus $2/29$ is stated to be the sum of $1/24$, $1/58$, $1/174$, and $1/232$. Probably Ahmes had no rule for forming the component fractions, and the answers given represent the accumulated experiences of previous writers. In one solitary case, however, he has indicated his method, for, after having asserted that $2/3$ is the sum of $1/2$ and $1/6$, he added that therefore two-thirds of one-fifth is equal to the sum of a half of a fifth and a sixth of a fifth, that is, to $1/10 + 1/30$.

That so much attention should have been paid to fractions may be explained by the fact that in early times their treatment presented considerable difficulty. The Egyptians and Greeks simplified the problem by reducing a fraction to the sum of several fractions, in each of which the numerator was unity, so that they had to consider only the various denominators: the sole exception to this rule being the fraction $2/3$. This remained the Greek practice until the sixth century of our era. The Romans, on the other hand, generally kept the denominator equal to twelve, expressing the fraction (approximately) as so many twelfths.

In Ahmes' treatment of multiplication, he seems to have relied on repeated additions. Thus, to multiply a certain number, which we will denote by the letter "a," by 13, he first multiplied by 2 and got $2a$, then he doubled the results and got $4a$, then he again doubled the result and got $8a$, and lastly he added together a, $4a$, and $8a$.

Now, we have used the sign "a" to stand for any number: not a particular number like 3, but *any* one. This is what Ahmes did, and what we learn to do in what we call "algebra." When Ahmes wished to find a number such that it, added to its seventh, makes 19, he symbolised the number by the sign we translate "heap." He had also signs for our "+," "−," and "=".[2] Nowadays we can write Ahmes' problem as: Find the number x

[2] In this book I shall take great care in distinguishing signs from what they signify. Thus 2 is to be distinguished from "2": by "2" I mean the *sign*, and the sign written without inverted commas indicates the thing signified. There has been, and is, much confusion, not only with beginners but with eminent mathematicians between a sign

such that $x + \frac{x}{7} = 19$. Ahmes gave the answer in the form $16 + \frac{1}{2} + \frac{1}{8}$.

We shall find that algebra was hardly touched by those Greeks who made of geometry such an important science, partly, perhaps, because the almost universal use of the abacus [3] rendered it easy for them to add and subtract without any knowledge of theoretical arithmetic. And here we must remember that the principal reason why Ahmes' arithmetical problems seem so easy to us is because of our use from childhood of the system of notation introduced into Europe by the Arabs, who originally obtained it from either the Greeks or the Hindoos. In this system an integral number is denoted by a succession of digits, each digit representing the product of that digit and a power of ten, and the number being equal to the sum of these products. Thus, by means of the local value attached to nine symbols and a symbol for zero, any number in the decimal scale of notation can be expressed. It is important to realise that the long and strenuous work of the most gifted minds was necessary to provide us with simple and expressive notation which, in nearly all parts of mathematics, enables even the less gifted of us to reproduce theorems which needed the greatest genius to discover. Each improvement in notation seems, to the uninitiated, but a small thing: and yet, in a calculation, the pen sometimes seems to be more intelligent than the user. Our notation is an instance of that great spirit of economy which spares waste of labour on what is already systematised, so that all our strength can be concentrated either upon what is known but unsystematised, or upon what is unknown.

Let us now consider the transformation of Egyptian geometry in Greek hands. Thales of Miletus (about 640–546 B.C.), who, during the early part of his life, was engaged partly in commerce and partly in public affairs, visited Egypt and first brought this knowledge into Greece. He discovered many things himself, and communicated the beginnings of many to his successors. We cannot form any exact idea as to how Thales presented his geometrical teaching. We infer, however, from Proclus that it consisted of a number of isolated propositions which were not arranged in a logical sequence, but that the proofs were deductive, so that the theorems were not a mere statement of an induction from a large number of special instances, as probably was the case with the Egyptian geometri-

---

and what is signified by it. Many have even maintained that *numbers* are the *signs* used to represent them. Often, for the sake of brevity, I shall use the word in inverted commas (say "*a*") as short for "what we call '*a*,'" but the context will make plain what is meant.

[3] The principle of the abacus is that a number is represented by counters in a series of grooves, or beads strung on parallel wires; as many counters being put on the first groove as there are units, as many on the second as there are tens, and so on. The rules to be followed in addition, subtraction, multiplication, and division are given in various old works on arithmetic.

cians. The deductive character which he thus gave to the science is his chief claim to distinction. Pythagoras (born about 580 B.C.) changed geometry into the form of an abstract science, regarding its principles in a purely abstract manner, and investigated its theorems from the immaterial and intellectual point of view. Among the successors of these men, the best known are Archytas of Tarentum (428–347 B.C.), Plato (429–348 B.C.), Hippocrates of Chios (born about 470 B.C.), Menaechmus (about 375–325 B.C.), Euclid (about 330–275 B.C.), Archimedes (287–212 B.C.), and Apollonius (260–200 B.C.).

The only geometry known to the Egyptian priests was that of surfaces, together with a sketch of that of solids, a geometry consisting of the knowledge of the areas contained by some simple plane and solid figures, which they had obtained by actual trial. Thales introduced the ideal of establishing by exact reasoning the *relations* between the different parts of a figure, so that some of them could be found by means of others in a manner strictly rigorous. This was a phenomenon quite new in the world, and due, in fact, to the abstract spirit of the Greeks. In connection with the new impulse given to geometry, there arose with Thales, moreover, scientific astronomy, also an abstract science, and undoubtedly a Greek creation. The astronomy of the Greeks differs from that of the Orientals in this respect: the astronomy of the latter, which is altogether concrete and empirical, consisted merely in determining the duration of some periods or in indicating, by means of a mechanical process, the motions of the sun and planets; whilst the astronomy of the Greeks aimed at the discovery of the geometrical laws of the motions of the heavenly bodies.

Let us consider a simple case. The area of a right-angled field of length 80 yards and breadth 50 yards is 4000 square yards. Other fields which are not rectangular can be approximately measured by mentally dissecting them—a process which often requires great ingenuity and is a familiar problem to land-surveyors. Now, let us suppose that we have a circular field to measure. Imagine from the centre of the circle a large number of radii drawn, and let each radius make equal angles with the next ones on each side of it. By joining the points in succession where the radii meet the circumference of the circle, we get a large number of triangles of equal area, and the sum of the areas of all these triangles gives an approximation to the area of the circle. It is particularly instructive repeatedly to go over this and the following examples mentally, noticing how helpful the abstract ideas we call "straight line," "circle," "radius," "angle," and so on, are. We all of us know them, recognise them, and can easily feel that they are trustworthy and accurate ideas. We feel at home, so to speak, with the idea of a square, say, and can at once give details about it which are *exactly* true for it, and *very nearly* true for a field which we know is *very nearly* a square. This replacement in thought by an abstract geometrical object economises labour of thinking and imagining by leading us to concentrate our thoughts on that alone which is essential for our purpose.

Thales seems to have discovered—and it is a good thing to follow these discoveries on figures made with the help of compasses and ruler—the proof of what may be regarded as the obvious fact that the circle is divided into halves by its diameter, that the angles at the base of a triangle with two equal sides—an "isosceles" triangle—are equal, that all the triangles described in a semi-circle with two of their angular points at the ends of the diameter and the third anywhere on the circumference contain a right angle, and he measured the distance of vessels from the shore, presumably by causing two observers at a known distance apart to measure the two angles formed by themselves and the ship. This last discovery is an application of the fact that a triangle is determined if its base and base angles are given.

When Archytas and Menaechmus employed mechanical instruments for solving certain geometrical problems, "Plato," says Plutarch, "inveighed against them with great indignation and persistence as destroying and perverting all the good there is in geometry; for the method absconds from incorporeal and intellectual to sensible things, and besides employs again such bodies as require much vulgar handicraft: in this way *mechanics* was dissimilated and expelled from geometry, and, being for a long time looked down upon by philosophy, became one of the arts of war." In fact, manual labour was looked down upon by the Greeks, and a sharp distinction was drawn between the slaves, who performed bodily work and really observed nature, and the leisured upper classes, who speculated and often only knew nature by hearsay. This explains much of the naïve, hazy, and dreamy character of ancient natural science. Only seldom did the impulse to make experiments for oneself break through; but when it did, a great progress resulted, as was the case with Archytas and Archimedes. Archimedes, like Plato, held that it was undesirable for a philosopher to seek to apply the results of science to any practical use; but, whatever might have been his view of what ought to be the case, he did actually introduce a large number of new inventions.

We will not consider further here the development of mathematics with other ancient nations, nor the chief problems investigated by the Greeks; such details may be found in some of the books mentioned in the Bibliography at the end. The object of this chapter is to indicate the nature of the science of geometry, and how certain practical needs gave rise to investigations in which appears an abstract science which was worthy of being cultivated for its own sake, and which incidentally gave rise to advantages of a practical nature.

There are two branches of mathematics which began to be cultivated by the Greeks, and which allow a connection to be formed between the spirits of ancient and modern mathematics.

The first is the method of geometrical analysis to which Plato seems to have directed attention. The analytical method of proof begins by assuming that the theorem or problem is solved, and thence deducing some result. If the result be false, the theorem is not true or the problem is incapable of solution: if the result be true, if the steps be reversible, we get (by reversing them) a synthetic proof; but if the steps be not reversible, no conclusion can be drawn. We notice that the leading thought in analysis is that which is fundamental in algebra, and which we have noticed in the case of Ahmes: the calculation or reasoning with an unknown entity, which is denoted by a conventional sign, as if it were known, and the deduction at last, of some relation which determines what the entity must be.

And this brings us to the second branch spoken of: algebra with the later Greeks. Diophantus of Alexandria, who probably lived in the early half of the fourth century after Christ, and probably was the original inventor of an algebra, used letters for unknown quantities in arithmetic and treated arithmetical problems analytically. Juxtaposition of symbols represented what we now write as "+," and "−" and "=" were also represented by symbols. All these symbols are mere abbreviations for words, and perhaps the most important advantage of symbolism—the power it gives of carrying out a complicated chain of reasoning almost mechanically—was not made much of by Diophantus. Here again we come across the economical value of symbolism: it prevents the wearisome expenditure of mental and bodily energy on those processes which can be carried out mechanically. We must remember that this economy both emphasises the unsubjugated—that is to say, unsystematised—problems of science, and has a charm—an æsthetic charm, it would seem—of its own.

Lastly, we must mention "incommensurables," "loci," and the beginnings of "trigonometry."

Pythagoras was, according to Eudemus and Proclus, the discoverer of "incommensurable quantities." Thus, he is said to have found that the diagonal and the side of a square are "incommensurable." Suppose, for example, that the side of the square is one unit in length; the diagonal is longer than this, but it is not two units in length. The excess of the length of the diagonal over one unit is not an integral submultiple of the unit. And, expressing the matter arithmetically, the remainder that is left over after each division of a remainder into the preceding divisor is not an integral submultiple of the remainder used as divisor. That is to say, the rule given in text-books on arithmetic and algebra for finding the greatest common measure does not come to an end. This rule, when applied to integer numbers, always comes to an end; but, when applied to certain lengths, it does not. Pythagoras proved, then, that if we start with a line of any length, there are other lines whose lengths do not bear to the first length the ratio of one integer to another, no matter if we have all the integers to choose from. Of course, any two fractions have the ratio of two integers to one another. In the above case of the diagonal, if the diagonal *were* in length some number x of units, we should have $x^2 = 2$, and it can be proved that no fraction, when "multiplied"—in the sense to be given in the next chapter—by itself gives 2 exactly, though there are fractions which give this result more and more approximately.

On this account, the Greeks drew a sharp distinction between "numbers," and "magnitudes" or "quantities" or measures of lengths. This distinction was gradually blotted out as people saw more and more the advantages of identifying numbers with the measures of lengths. The invention of analytical geometry, described in the third chapter, did most of the blotting out. It is in comparatively modern times that mathematicians have adequately realised the importance of this logically valid distinction made by the Greeks. It is a curious fact that the abandonment of strictly logical thinking should have led to results which transgressed what was then known of logic, but which are now known to be readily interpretable in the terms of what we now know of Logic. This subject will occupy us again in the sixth chapter.

The question of *loci* is connected with geometrical analysis, and is difficult to dissociate from a mental picture of a point in motion. Think of a point under restrictions to move only in some curve. Thus, a point may move so that its distance from a fixed point is constant; the peak of an angle may move so that the arms of the angle pass—slipping—through two fixed points, and the angle is always a right angle. In both cases the moving point keeps on the circumference of a certain circle. This curve is a "locus." It is evident how thinking of the locus a point can describe may help us to solve problems.

We have seen that Thales discovered that a triangle is determined if its base and base angles are given. When we have to make a survey of either an earthly country or part of the heavens, for the purpose of map-making, we have to measure angles—for example, by turning a *sight*, like those used on guns, through an angle measured in a circular arc of metal—to fix the relative directions of the stars or points on the earth. Now, for terrestrial measurements, a piece of country is approximately a flat surface, while the heavens are surveyed as if the stars were, as they seem to be, scattered on the inside of a sphere at whose centre we are. Secondly, it is a network of *triangles*—plane or spherical—of which we measure the angles and sometimes the sides: for, if the angles of a triangle are known, the *proportionality* of the sides is known; and this proportionality cannot be concluded from a knowledge of the angles of a rectangle, say. Hipparchus (born about 160 B.C) seems to have invented this practical science of the complete measurement of triangles from certain data. or, as it is called, "trigonometry," and the principles laid down by him were worked out by Ptolemy of Alexandria (died 168 A.D.) and also by the Hindoos and Arabians. Usually, only angles can be measured with accuracy, and so the question arises: given the magnitude of the angles, what can be concluded as to the kind of proportionality of the sides. Think of a circle described round the centre O, and let AP be the arc of this circle which measures the angle AOP. Notice that the ratio of AP to the radius is the same for the angle AOP whatever value the radius may have. Draw PM perpendicular to OA. Then the figure $OPMAP$ reminds one of a stretched bow, and hence are derived the names "sine of the arc AP" for the line PM, and "cosine" for OM. Tables of sines and cosines of arcs (or of angles, since the arc fixes the angle if the radius is known) were drawn up, and thus the sides PM and OM could be found in terms of the radius, when the arc was known. It is evident that this contains the essentials for the finding of the proportions of the sides of plane triangles. Spherical trigonometry contains more complicated relations which are directly relevant to the position of an astronomer and his measurements.

Mathematics did not progress in the hands of the Romans: perhaps the genius of this people was too practical. Still, it was through Rome that mathematics came into medieval Europe. The Arab mathematical textbooks and the Greek books from Arab translations were introduced into Western Europe by the Moors in the period 1150–1450, and by the end of the thirteenth century the Arabic arithmetic had been fairly introduced into Europe, and was practised by the side of the older arithmetic founded on the work of Boethius (about 475–526). Then came the Renascence. Mathematicians had barely assimilated the knowledge obtained from the Arabs, including their translations of Greek writers, when the refugees who escaped from Constantinople after the fall of the Eastern Empire (1453) brought the original works and the traditions of Greek science into Italy. Thus by the middle of the fifteenth century the chief results of Greek and Arabian mathematics were accessible to European students.

The invention of printing about that time rendered the dissemination of discoveries comparatively easy.

CHAPTER II

THE RISE AND PROGRESS OF MODERN MATHEMATICS—ALGEBRA

MODERN mathematics may be considered to have begun approximately with the seventeenth century. It is well known that the first 1500 years of the Christian era produced, in Western Europe at least, very little knowledge of value in science. The spirit of the Western Europeans showed itself to be different from that of the ancient Greeks, and only slightly less so from that of the more Easterly nations; and, when Western mathematics began to grow, we can trace clearly the historical beginnings of the use, in a not quite accurate form, of those conceptions—*variable* and *function*—which are characteristic of modern mathematics. We may say, in anticipation, that these conceptions, thoroughly analysed by reasoning as they are now, make up the difference of our modern views of Mathematics from, and have caused the likeness of them to, those of the ancient Greeks. The Greeks seem, in short, to have taken up a very similar position towards the mathematics of their day to that which logic forces us to take up towards the far more general mathematics of to-day. The generality of character has been attained by the effort to put mathematics more into touch with natural sciences—in particular the science of motion. The main difficulty was that, to reach this end, the way in which mathematicians expressed themselves was illegitimate. Hence philosophers, who lacked the real sympathy that must inspire all criticism that hopes to be relevant, never could discover any reason for thinking that what the mathematicians said was true, and the world had to wait until the mathematicians began logically to analyse their own conceptions. No body of men ever needed this sympathy more than the mathematicians from the revival of letters down to the middle of the nineteenth century, for no science was less logical than mathematics.

The ancient Greeks never used the conception of *motion* in their systematic works. The idea of a *locus* seems to imply that some curves could be thought of as generated by moving points; the Greeks discovered some things by helping their imaginations with imaginary moving points, but they never introduced the use of motion into their final proofs. This may have been because the Eleatic school, of which one of the principal representatives was Zeno (495–435 B.C.), invented some exceedingly subtle puzzles to emphasize the difficulty there is in the conception of motion. We shall return in some detail to these puzzles, which have not been appreciated in all the ages from the time of the Greeks till quite modern times. Owing to this lack of subtlety, the conception of variability was freely introduced into mathematics. It was the conceptions of *constant*,

variable, and ***function***, of which we shall, from now on, often have occasion to speak, which were generated by ideas of motion, and which, when they were logically purified, have made both modern mathematics and modern logic, to which they were transferred by mathematical logicians—Leibniz, Lambert, Boole, De Morgan, and the numerous successors of Boole and De Morgan from about 1850 onwards—into a science much more general than, but bearing some close analogies with, the ideal of Greek mathematical science. Later on will be found a discussion of what can be meant by a "moving point."

Let us now consider more closely the history of modern mathematics. Modern mathematics, like modern philosophy and like one part—the speculative and not the experimental part—of modern physical science, may be considered to begin with René Descartes (1596–1650). Of course, as we should expect, Descartes had many and worthy predecessors. Perhaps the greatest of them was the French mathematician François Viète (1540–1603), better known by his Latinized name of "Vieta." But it is simpler and shorter to confine our attention to Descartes.

Descartes always plumed himself on the independence of his ideas, the breach he made with the old ideas of the Aristotelians, and the great clearness and simplicity with which he described his ideas. But we must not underestimate the part that "ideas in the air" play; and, further, we know now that Descartes' breach with the old order of things was not as great as he thought.

Descartes, when describing the effect which his youthful studies had upon him when he came to reflect upon them, said:

"I was especially delighted with the mathematics, on account of the certitude and evidence of their reasonings: but I had not as yet a precise knowledge of their true use; and, thinking that they but contributed to the advancement of the mechanical arts, I was astonished that foundations so strong and solid should have had no loftier superstructure reared on them."

And again:

"Among the branches of philosophy, I had, at an earlier period, given some attention to logic, and, among those of the mathematics, to geometrical analysis and algebra—three arts or sciences which ought, as I conceived, to contribute something to my design. But, on examination, I found that, as for logic, its syllogisms and the majority of its other precepts are of avail rather in the communication of what we already know, or even in speaking without judgment of things of which we are ignorant, than in the investigation of the unknown: and although this science contains indeed a number of correct and very excellent precepts, there are, nevertheless, so many others, and these either injurious or superfluous, mingled with the former, that it is almost quite as difficult to effect a severance of the true from the false as it is to extract a Diana or a Minerva from a rough block of marble. Then as to the analysis of the ancients and the algebra of the moderns: besides that they embrace only matters highly abstract, and, to appearance, of no use, the former is so exclusively restricted to the consideration of figures that it can exercise the understanding only on condition of greatly fatiguing the imagination; and, in the latter, there is so complete a subjection to certain rules and formulas, that there results an art full of confusion and obscurity, calculated to embarrass, instead of a science fitted to cultivate the mind. By these considerations I was induced to seek some other method which would comprise the advantages of the three and be exempt from their defects. . . .

"The long chains of simple and easy reasonings by means of which geometers are accustomed to reach the conclusions of their most difficult demonstrations had led me to imagine that all things to the knowledge of which man is competent are mutually connected in the same way, and that there is nothing so far removed from us as to be beyond our reach, or so hidden that we cannot discover it, provided only that we abstain from accepting the false for the true, and always preserve in our thoughts the order necessary for the deduction of one truth from another. And I had little difficulty in determining the objects with which it was necessary to begin, for I was already persuaded that it must be with the simplest and easiest to know, and, considering that, of all those who have hitherto sought truth in the sciences, the mathematicians alone have been able to find any demonstrations, that is, any certain and evident reasons, I did not doubt but that such must have been the rule of their investigations. I resolved to begin, therefore, with the examination of the simplest objects, not anticipating, however, from this any other advantage than that to be found in accustoming my mind to the love and nourishment of truth and to a distaste for all such reasonings as were unsound. But I had no intention on that account of attempting to master all the particular sciences commonly denominated 'mathematics'; but observing that, however different their objects, they all agree in considering only the various relations or proportions subsisting among those objects, I thought it best for my purpose to consider these proportions in the most general form possible, without referring them to any objects in particular, except such as would most facilitate the knowledge of them, and without by any means restricting them to these, that afterwards I might thus be the better able to apply them to every other class of objects to which they are legitimately applicable. Perceiving further that, in order to understand these relations, I should have sometimes to consider them one by one and sometimes only to bear in mind or embrace them in the aggregate, I thought that, in order the better to consider them individually, I should view them as subsisting between straight lines, than which I could find no objects more simple or capable of being more distinctly represented to my imagination and senses; and, on the other hand, that, in order to retain them in the memory, or embrace an aggregate of many, I should express them by certain characters the briefest possible. In this way I believed that I could borrow all that was best both in geometrical analysis and in algebra, and correct all the defects of the one by help of the other."

Let us, then, consider the characteristics of algebra and geometry.

We have seen, when giving an account, in the first chapter, of the works of Ahmes and Diophantus, that mathematicians early saw the advantage of representing an unknown number by a letter or some other sign that may denote various numbers ambiguously, writing down—much as in geometrical analysis—the relations which they bear, by the conditions of the problem, to other numbers, and then considering these relations. If the problem is determinate—that is to say, if there are one or more definite solutions which can be proved to involve only numbers already fixed upon—this consideration leads, by the use of certain rules of calculation, to the determination—actual or approximate—of this solution or solutions. Under certain circumstances, even if there is a solution, depending on a variable, we can find it and express it in a quite general way, by rules, but that need not occupy us here.

Suppose that you know my age, but that I do not know yours, but wish to. You might say to me: "I was eight years old when you were born." Then I should think like this. Let x be the (unknown) number of years in your age at this moment and, say, 33 the number of years in my age at this moment; then in essentials your statement can be translated by the equation "$x - 8 = 33$." The meaning of the signs "$-$," "$=$," and "$+$" are supposed to be known—as indeed they are by most people nowadays quite sufficiently for our present purpose. Now, one of the rules of algebra is that any term can be taken from one side of the sign "$=$" to the other if only the "$+$" or "$-$" belonging to it is changed into "$-$" or "$+$," as the case may be. Thus, in the present case, we have: "$x = 33 + 8 = 41$." This absurdly simple case is chosen intentionally. It is essential in mathematics to remember that even apparently insignificant economies of thought add up to make a long and complicated calculation readily performed. This is the case, for example, with the convention introduced by Descartes of using the last letters of the alphabet to denote unknown numbers, and the first letters to denote known ones. This convention is adopted, with a few exceptional cases, by algebraists to-day, and saves much trouble in explaining and in looking for unknown and known quantities in an equation. Then, again, the signs "$+$," "$-$," "$=$" have great merits which those unused to long calculations cannot readily understand. Even the saving of space made by writing "xy" for "$x \times y$" ("x multiplied by y") is important, because we can obtain by it a shorter and more readily surveyed formula. Then, too, Descartes made a general practice of writing "powers" or "exponents" as we do now; thus "x^3" stands for "xxx" and "x^5" for some less suggestive symbol representing the continued multiplication of five x's.

One great advantage of this notation is that it makes the explanation of logarithms, which were the great and laborious discovery of John Napier (1550–1617), quite easy. We start from the equation "$x^m x^n = x^{m+n}$." Now, if $x^p = y$, and we call p the "logarithm of y to the base x"; in signs: "$p = \log_x y$"; the equation from which we started gives, if we denote x^m by "u" and x^n by "v," so that $m = \log_x u$ and $n = \log_x v$, that $\log_x (uv) = \log_x u + \log_x v$. Thus, if the logarithms of numbers to a given base (say $x = 10$) are tabulated, calculations with large numbers are made less arduous, for *addition replaces multiplication*, when logarithms are found. Also subtraction of logarithms gives the logarithm of the quotient of two numbers.

Let us now shortly consider the history of algebra from Diophantus to Descartes.

The word "algebra" is the European corruption of an Arabic phrase which means *restoration and reduction*—the first word referring to the fact that the same magnitude may be added to or subtracted from both sides of an equation, and the last word meaning the process of simplification. The science of algebra was brought among the Arabs by Mohammed ben Musa (Mahomet the son of Moses), better known as Alkarismi, in a work written about 830 A.D., and was certainly derived by him from the Hindoos. The algebra of Alkarismi holds a most important place in the history of mathematics, for we may say that the subsequent Arab and the early medieval works on algebra were founded on it, and also that through it the Arabic or Indian system of decimal numeration was introduced into the West. It seems that the Arabs were quick to appreciate the work of others—notably of the Greek masters and of the Hindoo mathematicians—but, like the ancient Chinese and Egyptians, they did not systematically develop a subject to any considerable extent.

Algebra was introduced into Italy in 1202 by Leonardo of Pisa (about 1175–1230) in a work based on Alkarismi's treatise, and into England by Robert Record (about 1510–1558) in a book called the *Whetstone of Witte* published in 1557. Improvements in the method or notations of algebra were made by Record, Albert Girard (1595–1632), Thomas Harriot (1560–1621), Descartes, and many others.

In arithmetic we use *symbols of number*. A symbol is any sign for a quantity, which is not the quantity itself. If a man counted his sheep by pebbles, the pebbles would be symbols of the sheep. At the present day, when most of us can read and write, we have acquired the convenient habit of using marks on paper, "1, 2, 3, 4," and so on, instead of such things as pebbles. Our "1 + 1" is abbreviated into "2," "2 + 1" is abbreviated into "3," "3 + 1" into "4," and so on. When "1," "2," "3," &c., are used to abbreviate, rather improperly, "1 mile," "2 miles," "3 miles," &c., for instance, they are called signs for *concrete* numbers. But when we shake off all idea of "1," "2," &c., meaning one, two, &c., of anything in particular, as when we say, "six and four make ten," then the numbers are called *abstract* numbers. To the latter the learner is first introduced in treatises on arithmetic, and does not always learn to distinguish rightly between the two. Of the operations of arithmetic only addition and subtraction can be performed with concrete numbers, and without speaking of more than one sort of 1. Miles can be added to miles, or taken from miles. Multiplication involves a new sort of 1, 2, 3, &c., standing for *repetitions* (or *times*, as they are called). Take 6 miles 5 times. Here are two kinds of units, 1 mile and 1 time. In multiplication, one of the units must be a number of repetitions or times, and to talk of multiplying 6 feet by 3 feet would be absurd. What notion can be formed of 6 feet taken "3 feet" times? In solving the following question, "If 1 yard cost 5 shillings, how much will 12 yards cost?" we do not multiply the 12 *yards* by the 5 *shillings*; the process we go through is the following: Since each yard costs 5 shillings, the buyer must put down 5 shillings as often (as many times) as the seller uses a one-yard measure; that is, 5 shillings is taken 12 times. In division we must have the idea either of repetition or of *partition*, that is, of cutting a quantity into a number of equal parts. "Divide 18 miles by 3 *miles*" means, find out how many *times* 3 miles must be repeated to give 18 miles: but "divide 18 miles by 3" means, cut 18 miles into 3 equal parts, and find how many miles are in each part.

The symbols of arithmetic have a *determinate connection*; for instance, 4 is always 2 + 2, whatever the things mentioned may be, miles, feet, acres, &c. In algebra we take symbols for numbers which have no determinate connection. As in arithmetic we draw conclusions about 1, 2, 3, &c., which are equally true of 1 foot, 2 feet, &c., 1 minute, 2 minutes, &c.; so in algebra we reason upon numbers in general, and draw conclusions which are equally true of all numbers. It is true that we also use, in kinds of algebra which have been developed within the last century, letters to represent things other than numbers—for example, *classes* of individuals with a certain property, such as "horned animals," for logical purposes; or certain geometrical or physical things with directions in space, such as "forces"—and signs like "+" and "−" to represent ways of combination of the things, which are analogous to, but not identical with, addition and subtraction. If "*a*" denotes "the class of horned animals" and "*b*" denotes "the class of beasts of burden," the sign "*ab*" has been used to denote "the class of horned beasts of burden." We see that here $ab = ba$, just as in the multiplication of numbers, and the above operation has been called, partly for this reason, "logical multiplication," and denoted in the above way. Here we meet the practice of mathematicians—and of all scientific men—of using words in a wider sense for the sake of some analogy. This habit is all the more puzzling to many people because mathematicians are often not conscious that they do it, or even talk sometimes as if they thought that they were generalising *conceptions* instead of words. But, when we talk of a "family tree," we do not indicate a widening of our conception of trees of the roadside.

We shall not need to consider these modern algebras, but we shall be constantly meeting what are called the "generalisations of number" and transference of methods to analogous cases. Indeed, it is hardly too much to say that in this lies the very spirit of discovery. An example of this is given by the extension of the word "numbers" to include the names of *fractions as well.* The occasion for this extension was given by the use of arithmetic to express such quantities as distances. This had been done by Archimedes and many others, and had become the usual method of procedure in the works of the mathematicians of the sixteenth century, and plays a great part in Descartes' work.

Mathematicians, ever since they began to apply arithmetic to geometry, became alive to the fact that it was convenient to represent points on a straight line by numbers, and numbers by points on a straight line. What is meant by this may be described as follows. If we choose a unit of length, we can mark off points in a straight line corresponding to 0 units —which means that we select a point, called "the origin," to start from,— 1 unit, 2 units, 3 units, and so on, so that "the point *m*," as we will call it for short, is at a distance of *m* units from the origin. Then we can divide up the line and mark points corresponding to the fractions $1/2$, $2/3$, $7/8$, $1/13$, $5/3$, or the point between 1 and 2 which is the same distance from 1 as $2/3$ is from 0, and so on. Now, there is nothing here to distinguish fractions from numbers. Both are treated exactly in the same way; the results of addition, subtraction, multiplication, and division [4] are interpretable, in much the same way as new points whether the "*a*" and "*b*" in

[4] The operation of what is called, for the sake of analogy, "multiplication" of fractions is defined in the manner indicated in the following example. If $3/4$ of a yard costs 10d., how much does $7/8$ of a yard cost? The answer is $\frac{10 \times 4 \times 7}{3 \times 8}$ pence, and we define $\frac{4 \times 7}{3 \times 8}$ as $\frac{1}{3/4}$ "multiplied by" $7/8$, by analogy with what would happen if $3/4$ were 1 and $7/8$ were, say, 3.

"$a + b$," "$a - b$," "ab," and so on, stand for numbers or fractions, and we have, for example,

$$a + b = b + a,\ ab = ba,\ a\,(b + c) = ab + ac,$$

always. Because of this very strong analogy, mathematicians have called the fractions "numbers" too, and they often speak and write of "generalisations of numbers," of which this is the first example, as if the conception of number were generalised, and not merely the *name* "number," in virtue of a great and close and important analogy.

When once the points of a line were made to represent numbers, there seemed to be no further difficulty in admitting certain "irrational numbers" to correspond to the end-points of the incommensurable lines which had been discovered by the Greeks. This question will come up again at a later stage: there are necessary discussions of principle involved, but mathematicians did not go at all deeply into questions of principle until fairly modern times. Thus it has happened that, until the last sixty years or so, mathematicians were nearly all bad reasoners, as Swift remarked of the mathematicians of Laputa in *Gulliver's Travels*, and were unpardonably hazy about first principles. Often they appealed to a sort of faith. To an intelligent and doubting beginner, an eminent French mathematician of the eighteenth century said: "Go on, and faith will come to you." It is a curious fact that mathematicians have so often arrived at truth by a sort of instinct.

Let us now return to our numerical algebra. Take, say, the number 8, and the fraction, which we will now call a "number" also, ⅛. Add 1 to both; the greater contains the less exactly 8 times. Now this property is possessed by *any* number, and not 8 alone. In fact, if we denote the number we start with by "a," we have, by the rules of algebra, $\dfrac{a+1}{{}^1/_a+1} = a.$ This is an instance of a general property of numbers proved by algebra.

Algebra contains many rules by which a complicated algebraical expression can be reduced to its simplest terms. Owing to the suggestive and compact notation, we can easily acquire an almost mechanical dexterity in dealing with algebraical symbols. This is what Descartes means when he speaks of algebra as not being a science fitted to cultivate the mind. On the other hand, this art is due to the principle of the economy of thought, and the mechanical aspect becomes, as Descartes foresaw, very valuable if we could use it to solve geometrical problems without the necessity of fatiguing our imaginations by long reasonings on geometrical figures.

I have already mentioned that the valuable notation "x^m" was due to Descartes. This was published, along with all his other improvements in algebra, in the third part of his *Geometry* of 1637. I shall speak in the next chapter of the great discovery contained in the first two parts of this work; here I will resume the improvements in notation and method made by Descartes and his predecessors, which make the algebraical part of the *Geometry* very like a modern book on algebra.

It is still the custom in arithmetic to indicate addition by juxtaposition: thus "2½" means "2 + ½." In algebra, we always, nowadays, indicate addition by the sign "+" and multiplication by juxtaposition or, more rarely, by putting a dot or the sign "×" between the signs of the numbers to be multiplied. Subtraction is indicated by "–".

Here we must digress to point out—what is often, owing to confusion of thought, denied in text-books—that, where "a" and "b" denote numbers, "$a - b$" can only denote a number if a is equal to or greater than b. If a is equal to b, the number denoted is zero; there is really no good reason for denying, say, that the numbers of Charles II.'s foolish sayings and wise deeds are equal, if a well-known epitaph be true. Here again we meet the strange way in which mathematics has developed. For centuries mathematicians used "negative" and "positive" numbers, and identified "positive" numbers with signless numbers like 1, 2, and 3, without any scruple, just as they used fractionary and irrational "numbers." And when logically-minded men objected to these wrong statements, mathematicians simply ignored them or said: "Go on; faith will come to you." And the mathematicians were right, and merely could not give correct reasons—or at least always gave wrong ones—for what they did. We have, over again, the fact that criticism of the mathematicians' procedure, if it wishes to be relevant, must be based on thorough sympathy and understanding. It must try to account for the rightness of mathematical views, and bring them into conformity with logic. Mathematicians themselves never found a competent philosophical interpreter, and so nearly all the interesting part of mathematics was left in obscurity until, in the latter half of the nineteenth century, mathematicians themselves began to cultivate philosophy—or rather logic.

Thus we must go out of the historical order to explain what "negative numbers" means. First, we must premise that when an algebraical expression is enclosed in brackets, it signifies that the whole result of that expression stands in the same relation to surrounding symbols as if it were one letter only. Thus, "$a - (b - c)$" means that from a we are to take $b - c$, or what is left after taking c from b. It is not, therefore, the same as $a - b - c$. In fact we easily find that $a - (b - c)$ is the same as $a - b + c$. Note also that "$(a + b)\ (c + d)$" means $(a + b)$ multiplied by $(c + d)$.

Now, suppose a and b are numbers, and a is greater than b. Let $a - b$ be c. To get c from a, we carry out the operation of taking away b. *This operation, which is the fulfilment of the order: "Subtract b," is a "negative number."* Mathematicians call it a "number" and denote it by "$-b$" simply because of analogy: the same rules for calculation hold for "negative numbers" and "positive numbers" like "$+b$," whose meaning is now clear too, as do for our signless numbers; when "addition," "subtraction," &c., are redefined for these operations. The way in which this redefinition must take place is evident when we represent integers, fractions, and positive and negative numbers by points on a straight line. To the right of 0 are the integers and fractions, to the left of 0 are the negative numbers, and to the right of 0 stretch the series of positive numbers, $+a$ coinciding with a and being symmetrically placed with $-a$ as regards 0. Also we determine that the operations of what we call "addition," &c., of these new "numbers" *must lead to the same results* as the former operations of the same name. Thus the same symbol is used in different senses, and we write

$$a + b - b = a + 0 = (+a) + (+b) + (-b) = +a = a.$$

This is a remarkable sequence of quick changes.

We have used the sign of equality, "=". It means originally, "is the same as." Thus $3 + 1 = 4$. But we write, by the above convention, "$a = +a$," and so we sacrifice exactness, which sometimes looks rather pedantic, for the sake of keeping our analogy in view, and for brevity.

Let us bear this, at first sight, puzzling but, at second sight, justifiable peculiarity of mathematicians in mind. It has always puzzled intelligent beginners and philosophers. The laws of calculation and convenient symbolism are *the* things a mathematician thinks of and aims at. He seems to identify different things if they both satisfy the same laws which are important to him, just as a magistrate may think that there is not much difference between Mr. A., who is red-haired and a tinker and goes to chapel, and Mr. B., who is a brown-haired horse-dealer and goes to church, if both have been found out committing petty larceny. But their respective ministers of religion or wives may still be able to distinguish them.

Any two expressions connected by the sign of equality form an "equation." Here we must notice that the words "Solve the equation $x^2 + ax = b$" mean: find the value or values of x such that, a and b being given numbers, $x^2 + ax$ becomes b. Thus, if $a = 2$ and $b = -1$, the solution is $x = -1$.

As we saw above, Descartes fixed the custom of employing the letters at the beginning of the alphabet to denote known quantities, and those at the end of the alphabet to denote unknown quantities. Thus, in the above example, a and b are some numbers supposed to be given, while x is sought. The question is solved when x is found in terms of a and b and fixed numbers (like 1, 2, 3); and so, when to a and b are attributed any fixed values, x becomes fixed. The signs "a" and "b" denote ambiguously, not uniquely like "2" does; and "x" does not always denote ambiguously when a and b are fixed. Thus, in the above case, when $a = 2$, $b = -1$, "x"

denotes the one negative number -1. What is meant is this: In each member of the class of problems got by giving a and b fixed values independently of one another, there is an unknown x, which may or may not denote different numbers, which only becomes known when the equation is solved. Consider now the equation $ax + by = c$, where a, b, and c are known quantities and x and y are unknown. We can find x in terms of a, b, c, and y, or y in terms of a, b, c, and x; but x is only fixed when y is fixed, or y when x is fixed. Here in each case of fixedness of a, b, and c, x is undetermined and "variable," that is to say, it may take any of a whole class of values. Corresponding to each x, one y belongs; and y also is a "variable" depending on the "independent variable" x. The idea of "variability" will be further illustrated in the next chapter; here we will only point out how the notion of what is called by mathematicians the "functional dependence" of y on x comes in. The variable y is said to be a "function" of the variable x if to every value of x corresponds one or more values of y. This use has, to some extent, been adopted in ordinary language. We should be understood if we were to say that the amount of work performed by a horse is a function of the food that he eats.

Descartes also adopted the custom—if he did not arrive at it independently—advocated by Harriot of transferring all the terms of an equation to the same side of the sign of equality. Thus, instead of "$x = 1$," "$ax + b = c$," and "$3x^2 + g = hx$," we write respectively "$x - 1 = 0$," "$ax + (b - c) = 0$," and "$3x^2 - hx + g = 0$." The point of this is that all equations of the same degree in the unknown—we shall have to consider cases of more unknowns than one in the next chapter—that is to say, equations in which the highest power of x (x or x^2 or x^3 . . .) is the same, are easily recognisable. Further, it is convenient to be able to speak of the expression which is equated to 0 as well as of the equation. The equations in which x^2, and no higher power of x, appears are called "quadratic" equations—the result of equating a "quadratic" function to 0; those in which x^3, and no higher power, appears are called "cubic"; and so on for equations "of the fourth, fifth . . ." degrees. Now the quadratic equations, $3x^2 + g = 0$, $ax^2 + bx + c = 0$, $x^2 - 1 = 0$, for example, are different, but the differences are unimportant in comparison with this common property of being of the same degree: all can be solved by modifications of one general method.

Here it is convenient again to depart from the historical order and briefly consider the meaning of what are called "imaginary" expressions. If we are given the equation $x^2 - 1 = 0$, its solutions are evidently $x = +1$ or $x = -1$, for the square roots of $+1$ are $+1$ and -1. But if we are given the equation $x^2 + 1 = 0$, analogy would lead us to write down the two solutions $x = +\sqrt{-1}$ and $x = -\sqrt{-1}$. But there is no positive or negative "number" which we have yet come across which, when multiplied by itself, gives a negative "number." Thus "imaginary numbers" were rejected by Descartes and his followers. Thus $x^2 - 1 = 0$ had two solutions, but $x^2 + 1 = 0$ none; further, $x^3 + x^2 + x + 1 = 0$ had one solution ($x = -1$), while $x^3 - x^2 - x + 1 = 0$ had two ($x = 1$, $x = -1$), and $x^3 - 2x^2 - x + 2 = 0$ had three ($x = 1$, $x = -1$, $x = 2$). Now, suppose, for a moment, that we can have "imaginary" roots and $(\sqrt{-1})(\sqrt{-1}) = -1$, and also that we can speak of *two* roots when the roots are identical in a case like the equation $x^2 + 2x + 1 = 0$, or $(x + 1)^2 = 0$, which has two identical roots $x = -1$. Then, in the above five equations, the first two quadratic ones have two roots each ($+1$, -1, and $+\sqrt{-1}$, $-\sqrt{-1}$ respectively), and the three cubics have three each (-1, $+\sqrt{-1}$, $-\sqrt{-1}$; $+1$, -1, $+1$; and $+1$, -1, $+2$ respectively). In the general case, the theorem has been proved that every equation has as many roots as (and not merely "no more than," as Descartes said) its degree has units. For this and for many other reasons like it in enabling theorems to be stated more generally, "imaginary numbers" came to be used almost universally. This was greatly helped by one puzzling circumstance: true theorems can be discovered by a process of calculation with imaginaries. The case is analogous to that which led mathematicians to introduce and calculate with "negative numbers."

For the case of imaginaries, let a, b, c, and d be any numbers, then

$$\begin{aligned}(a^2 + b^2)(c^2 + d^2) &= (a + b\sqrt{-1})(a - b\sqrt{-1}) \\ &\quad (c + d\sqrt{-1})(c - d\sqrt{-1}) \\ &= (a + b\sqrt{-1})(c + d\sqrt{-1}) \\ &\quad (a - b\sqrt{-1})(c - d\sqrt{-1}) \\ &= [(ac - bd) + \sqrt{-1}(ad + bc)] \\ &\quad [(ac - bd) - \sqrt{-1}(ad + bc)] \\ &= (ac - bd)^2 + (ad + bc)^2.\end{aligned}$$

We get, then, an interesting and easily verifiable theorem on numbers by calculation with imaginaries, and imaginaries disappear from the conclusion. Mathematicians thought, then, that imaginaries, though apparently uninterpretable and even self-contradictory, *must* have a logic. So they were used with a faith that was almost firm and was only justified much later. Mathematicians indicated their growing security in the use of $\sqrt{-1}$ by writing "*i*" instead of "$\sqrt{-1}$" and calling it "the complex unity," thus denying, by implication, that there is anything really imaginary or impossible or absurd about it.

The truth is that "*i*" is not uninterpretable. It represents an operation, just as the negative numbers do, but is of a different kind. It is geometrically interpretable also, though not in a straight line, but in a plane. For this we must refer to the Bibliography; but here we must point out that, in this "generalisation of number" again, the words "addition," "multiplication," and so on, do not have exactly the same, but an analogous, meaning to those which they had before, and that "complex numbers" form a domain like a plane in which a line representing the integers, fractions, and irrationals is contained. But we must leave the further development of these questions.

It must be realised that the essence of algebra is its generality. In the most general case, every symbol and every statement of a proposition in algebra is interpretable in terms of certain operations to be undertaken with abstract things such as numbers or classes or propositions. These operations merely express the relations of these things to one another. If the results at any stage of an algebraical process can be interpreted—and this interpretation is often suggested by the symbolism—say, not as operations with operations with integers, but as other operations with integers, they express true propositions. Thus $(a + b)^2 = a^2 + 2ab + b^2$ expresses, for example, a relation holding between those operations with integers that we call "fractionary numbers," or an analogous relation between integers. The language of algebra is a wonderful instrument for expressing shortly, perspicuously, and suggestively, the exceedingly complicated relations in which abstract things stand to one another. The motive for studying such relations was originally, and is still in many cases, the close analogy of relations between certain abstract things to relations between certain things we see, hear, and touch in the world of actuality round us, and our minds are helped in discovering such analogies by the beautiful picture of algebraical processes made in space of two or of three dimensions made by the "analytical geometry" of Descartes, described in the next chapter.

CHAPTER III

THE RISE AND PROGRESS OF MODERN MATHEMATICS—ANALYTICAL GEOMETRY AND THE METHOD OF INDIVISIBLES

WE will now return to the consideration of the first two sections of Descartes' book *Geometry* of 1637.

In Descartes' book we have to glean here and there what we now recognise as the essential points in his new method of treating geometrical questions. These points were not expressly stated by him. I shall, however, try to state them in a small compass.

Imagine a curve drawn on a plane surface. This curve may be considered as a *picture* of an algebraical equation involving x and y in the following way. Choose any point on the curve, and call "x" and "y" the numbers that express the perpendicular distances of this point, in terms of a unit of length, from two straight lines (called "axes") drawn at right angles to one another in the plane mentioned. Now, as we move from point to point of the curve, x and y both vary, *but there is an unvarying relation which connects x and y, and this relation can be expressed by an algebraical equation called "the equation of the curve," and which contains, in germ as it were, all the properties of the curve considered.* This constant relation between x and y is a relation like $y^2 = 4ax$. We must distinguish carefully between a constant relation between variables and a relation between constants. We are always coming across the former kind of relation in mathematics; we call such a relation a "function" of x and y —the word was first used about fifty years after Descartes' *Geometry* was published, by Leibniz—and write a function of x and y in general as "$f(x, y)$." In this notation, no hint is given as to any particular relation x and y may bear to each other, and, in such a particular function as $y^2 - 4ax$, we say that "the *form* of the function is constant," and this is only another way of saying that the relation between x and y is fixed. This may be also explained as follows. If x is fixed, there is fixed one or more values of y, and if y is fixed, there is fixed one or more values of x. Thus the equation $ax + by + c = 0$ gives one y for each x and one x for each y; the equation $y^2 - 4ax = 0$ gives two y's for each x and one x for each y.[5]

Consider the equation $ax + by + c = 0$, or, say, the more definite instance $x + 2y - 2 = 0$. Draw axes and mark off points; having fixed on a unit of length, find the point $x = 1$ on the x-axis, on the perpendicular to this axis measure where the corresponding y, got by substituting $x = 1$ in the above equation, brings us. We find $y = \frac{1}{2}$. Take $x = \frac{1}{3}$, then $y = \frac{5}{6}$; and so on. We find that all the points on the parallels to the y-axis lie on one straight line. This straight line is determined by the equation $x + 2y - 2 = 0$; every point off that straight line is such that its x and y are not connected by the relation $x + 2y - 2 = 0$, and every point of it is such that its x and y are connected by the relation $x + 2y - 2 = 0$. Similarly we can satisfy ourselves that every point on the circumference of a circle of radius c units of length, described round the point where the axes cross, is such that $x^2 + y^2 = c^2$, and every point not on this circumference does not have an x and y such that the constant relation $x^2 + y^2 = c^2$ is satisfied for it.

There are two points to be noticed in the above general statement. Firstly, I have said that the curve "*may* be expressed," and so on. By this I mean that it is possible—and not necessarily always true—that the curve

[5] We also denote a function of x by "$f(x)$" or "$F(x)$" or "$\phi(x)$," &c. Here "f" is a sign for "function of," not for a number, just as later we shall find "sin" and "Δ" and "d" standing for functions and not numbers. This may be regarded as an extension of the language of early algebra. The equation $y = f(x)$ is in a good form for graphical representation in the manner explained below.

may be so considered. We can imagine curves that cannot be represented by a finite algebraical equation. Secondly, about the fundamental lines of reference—the "axes" as they are called. One of these axes we have called the "x-axis," and the distance measured by the number x is sometimes called "the abscissa"; while the line of length y units which is perpendicular to the end of the abscissa farthest from the origin, and therefore parallel to the other axis ("the y-axis") is called "the ordinate." The name "ordinate" was used by the ancient Roman surveyors. The lines measured by the numbers x and y are called the "co-ordinates" of the point determining and determined by them. Sometimes the numbers x and y themselves are called "co-ordinates," and we will adopt that practice here.

Sometimes the axes are not chosen at right angles to one another, but it is nearly always far simpler to do so, and in this book we always assume that the axes are rectangular. The whole plane is divided by the axes into four partitions, the co-ordinates are measured from the point—called "the origin"—where the axes cross. Here the interpretation in geometry of the "negative quantities" of algebra—which so often seems so puzzling to intelligent beginners—gives us a means of avoiding the ambiguity arising from the fact that there would be a point with the same co-ordinates in each quadrant into which the plane is divided.

Consider the x-axis. Measure lengths on it from the origin, so that to the origin (O) corresponds the number 0. Let OA, measured from left to right along the axis, be the unit of length; then to the point A corresponds the number 1. Then let lengths AB, BC, and so on, all measured from left to right, be equal to OA in length; to the points B, C, and so on, correspond the numbers 2, 3, and so on. Further to the point that bisects OA let the fraction $\frac{1}{2}$ correspond; and so on for the other fractions. In this way half of the x-axis is nearly filled up with points. But there are points, such as the point P, such that OP is the length of the circumference of a circle, say of unit diameter. For picturesqueness, we may imagine this point P got by rolling the circle along the x-axis from O through one revolution. The point P will fall a little to the left of the point $3\frac{1}{7}$ and a little to the right of the point $3\frac{7}{50}$, and so on; the point P is not one of the points to which names of fractions have been assigned by the process sketched above. This can be proved rigidly. If it were not true, it would be very easy to "square the circle."

There are many other points like this. There is no fraction which, multiplied by itself, gives 2; but there is a length—the diagonal of a square of unit side—which is such that, if *we were to assume that a number corresponded to every point on OX*, it would be a number a such that $a^2 = 2$. We will return to this important question of the correspondence of points and lines to numbers, and will now briefly recall that "negative numbers" are represented, in Descartes' analytical geometry, on the x-axis, by the points to the left of the origin, and, on the y-axis, by the points below the origin. This was explained in the second chapter.

Algebraical geometry gave us a means of classifying curves. All straight lines determine equations of the first degree between x and y, and all such equations determine straight lines; all equations of the second degree between x and y, that is to say, of the form

$$ax^2 + bxy + cy^2 + dx + ey + f = 0,$$

determine curves which the ancient Greeks had studied and which result from cutting a solid circular cone, or two equal cones with the same axis, whose only point of contact is formed by the vertices. It is somewhat of a mystery why the Greek geometricians should have pitched upon these particular curves to study, and we can only say that it seems, from the present standpoint, an exceedingly lucky chance. For these "conic sections"—of which, of course, the circle is a particular case—are all the curves, and those only, which are represented by the above equation of the second degree. The three great types of curves—the "parabola," the "ellipse," and the "hyperbola"—all result from the above equation when the coefficients a, b, c, d, e, f satisfy certain special conditions. Thus, the equation of a circle—which is a particular kind of ellipse—is always of the form got from the above equation by putting $b = 0$ and $c = a$.

It may be mentioned that, long after these curves were introduced as sections of a cone, Pappus discovered that they could all be defined in a plane as loci of a point P which moves so that the proportion that the distance of P from a fixed point (S) bears to the perpendicular distance of P (PN) to a fixed straight line is constant. As this proportion is less than equal to or greater than 1, the curve is an ellipse, parabola, or hyperbola, respectively.

It will not be expected that a detailed account should here be given of the curves which result from the development of equations of the second or higher degrees between x and y. I will merely again emphasize some points which are, in part, usually neglected or not clearly stated in text-books. The letters "$a, b, \ldots x, y$," here stand for "numbers" in the extended sense. We have seen in what sense we may, with the mathematicians, speak of fractionary, positive, and negative "numbers," and identify, say, the positive number $+2$ and the fraction $2/1$ with the signless integer 2. Well, then, the above letters stand for numbers of that class which includes in this sense the fractionary, irrational, positive and negative numbers, but excludes the imaginary numbers. We call the numbers of this class "real" numbers. The question of irrational numbers will be discussed at greater length in the sixth chapter, but enough has been said to show how they were introduced. In mathematics it has, I think, always happened that conceptions have been used long before they were formally introduced, and used long before this use could be logically justified or its nature clearly explained. The history of mathematics is the history of a faith whose justification has been long delayed, and perhaps is not accomplished even now.

These numbers are the measurements of length, in terms of a definite unit, like the inch, of the abscissæ and ordinates of certain points. We speak of such points simply by naming their co-ordinates, and say, for example, that "the distance of the point (x, y) from the point (a, b) is the positive square root of $(x - a)^2 + (y - b)^2$."

Notice that x^2, for example, is the length of a *line*. It is natural to make, as algebraists before Descartes did, x^2 stand *primarily* for the number of square units in a square whose sides are x units in length, but there is no necessity in this. We shall often use the latter kind of measurement in the fourth and fifth chapters.

The equation of a straight line can be made to satisfy two given conditions. We can write the equation in the form

$$x + \frac{b}{a} y + \frac{c}{a} = 0,$$

and thus have two ratios, $\frac{b}{a}$ and $\frac{c}{a}$, that we can determine according to the conditions. The equation $ax + by + c = 0$ has apparently *three* "arbitrary constants," as they are called, but we see that this greater generality is only apparent. Now we can so fix these constants that two conditions are fulfilled by the straight line in question. Thus, suppose that one of these conditions is that the straight line should pass through the origin—the point $(0, 0)$. This means simply that when $x = 0$, then $y = 0$. Putting them, $x = 0$ and $y = 0$ in the above equation, we get $\frac{c}{a} = 0$, and thus one of the constants is determined. The other is determined by a new condition that, say, the line also passes through the point $(1/3, 2)$. Substituting, then, in the above equation, we have, as $\frac{c}{a} = 0$ as we know already, $1/3 + \frac{2b}{a} = 0$, whence $\frac{b}{a} = -1/6$. Hence the equation of the line passing through $(0, 0)$ and $(1/3, 2)$ is $x - 1/6\, y = 0$, or $y - 6x = 0$. Instead of having to pass through a certain point, a condition may be, for example, that the perpendicular from the origin on the straight line should be of a certain length, or that the line should make a certain angle with the x-axis, and so on.

Similarly, the circle whose equation is written in the form

$$(x - a)^2 + (y - b)^2 = c^2$$

is of radius c and centre (a, b). It can be determined to pass through any *three* points, or, say, to have a determined length of radius and position of centre. Fixation of centre is equivalent to two conditions. Thus, suppose the radius is to be of unit length: the above equation is $(x - a)^2 + (y - b)^2 = 1$. Then, if the centre is to be the origin, both a and b are determined to be 0, and this may also be effected by determining that the circle is to pass through the points $(\frac{1}{2}, 0)$ and $(-\frac{1}{2}, 0)$, for example.

Now, if we are to find the points of intersection of the straight line $2x + 2y = 1$ and the circle $x^2 + y^2 = 1$, we seek those points which are common to both curves, that is to say, all the pairs of values of x and y which satisfy *both* the above equations. Thus we need not trouble about the geometrical picture, but we only have to apply the rules of algebra for finding the values of x and y which satisfy two "simultaneous" equations in x and y. In the above case, if (X, Y) is a point of intersection, we have $Y = \dfrac{1 - 2X}{2}$, and therefore, by substitution in the other equation, $X^2 + \left(\dfrac{1 - 2X}{2}\right)^2 = 1$. This gives a quadratic equation

$$8X^2 - 4X - 3 = 0$$

for X, and, by rules, we find that X must be either $\frac{1}{4}(1 + \sqrt{7})$ or $\frac{1}{4}(1 - \sqrt{7})$. Hence there are *two* values of the abscissa which are given when we ask what are the co-ordinates of the points of intersection; and the value of y which corresponds to each of these x's is given by substitution in the equation $2x + 2y = 1$.

Thus we find again the fact, obvious from a figure, that a straight line cuts [a circle] at two points at most. We can determine the points of intersection of any two curves whose equations can be expressed algebraically, but of course the process is much more complicated in more general cases. Here we will consider an important case of intersection of a straight line.

Think of a straight line cutting a circle at two points. Imagine one point fixed and the other point moved up towards the first. The intersecting line approaches more and more to the position of the tangent to the circle at the first point, and, by making the movable point approach the other closely enough, the secant will approach the tangent in position as nearly as we wish. Now, a tangent to a curve at a certain point was defined by the Greeks as a straight line through the point such that between it and the curve no other ***straight line*** could be drawn. Note that other ***curves*** might be drawn: thus various circles may have the same tangent at a common point on their circumference, but no circle—and no curve met with in elementary mathematics—has more than one tangent at a point. Descartes and many of his followers adopted different forms of definition which really involve the idea of a *limit,* an idea which appears boldly in the infinitesimal calculus. A tangent is the *limit* of a secant as the points of intersection approach infinitely near to one another; it is a produced side of the polygon with infinitesimal sides that the curve is supposed to be; it is the direction of motion at an instant of a point moving in the curve considered. The equation got from that of the curve by substituting for y from the equation of the intersecting straight line has, if this straight line is a tangent, two equal roots. In the above case, this equation was quadratic. In the case of a circle, we can easily deduce the well-known property of a tangent of being perpendicular to the radius; and see that this property has no analogue in the case of other curves.

We must remember that, just as *plane* curves determine and are determined by equations with *two* independent variables x and y, so surfaces—spheres, for instance—in three-dimensional space determine and are determined by equations with *three* independent variables, x, y, and z. Here x, y, and z are the co-ordinates of a point in space; that is to say, the numerical measures of the distances of this point from three fixed planes at right angles to each other. Thus, the equation of a sphere of radius d and centre at (a, b, c) is $(x-a)^2+(y-b)^2+(z-c)^2=d^2$.

We may look at analytical geometry from another point of view which we shall find afterwards to be important, and which even now will suggest to us some interesting thoughts. The essence of Descartes' method also appears when we represent *loci* by the method. Consider a circle; it is the locus of a point (P) which moves in a plane so as to preserve a constant distance from a fixed point (O). Here we may think of P as varying in position, and make up a very striking picture of what we call a *variable* in mathematics. We must, however, remember that, by what we call a "variable" for the sake of picturesqueness, we do not necessarily mean something which varies. Think of the point of a pen as it moves over a sheet of writing paper; it occupies different positions with respect to the paper at different times, and we understandably say that the pen's point moves. But now think of a point in space. A geometrical point—which is not the bit of space occupied by the end of a pen or even an "atom" of matter—is merely a mark of position. We cannot, then, speak of a point moving; the very essence of point is to *be* position. The motion of a point of *space,* as distinguished from a point of matter, is a fiction, and is the supposition that a given point can now be one point and now another. Motion, in the ordinary sense, is only possible to matter and not to space.

Thus, when we speak of a "variable position," we are speaking absurdly if we wish our words to be taken literally. But we do not really so wish when we come to think about it. What we are doing is this: we are using a picturesque phrase for the purpose of calling up an easily imagined thought which helps us to visualise roughly a mathematical proposition which can only be described accurately by a prolix process. The ancient Greeks allowed prolixity, and it was only objected to by the uninitiated. Modern mathematics up to about sixty years ago successfully warred against prolixity; hence the obscurity of its fundamental notions and processes and its great conquests. The great conquests were made by sacrificing very much to analogy: thus, entities like the integer 2, the ratio 2/1, and the real number which is denoted by "2" were identified, as we have seen, because of certain close analogies that they have. This seems to have been the chief reason why the procedure of the mathematicians has been so often condemned by logicians and even by philosophers. In fact, when mathematicians began to try to find out the nature of Mathematics, they had to examine their entities and the methods which they used to deal with them with the minutest care, and hence to look out for the points when the analogies referred to break down, and distinguish between what mathematicians had usually failed to distinguish. Then the people who do not mind a bit what Mathematics *is,* and are only interested in what it *does,* called these earnest inquirers "pedants" when they should have said "philosophers," and "logic-choppers"—whatever they may be—when they should have said "logicians." We have tried to show why ratios or fractions, and so on, are called "numbers," and apparently said to be something which they are not; we must now try to get at the meaning of the words "constant" and "variable."

By means of algebraic formulæ, rules for the reconstruction of great numbers—sometimes an infinity—of facts of nature may be expressed very concisely or even embodied in a *single* expression. The essence of the formula is that it is an expression of a *constant* rule among *variable* quantities. These expressions "constant" and "variable" have come down into ordinary language. We say that the number of miles which a certain man walks per day is a "variable quantity"; and we do not mean that, on a particular day, the number was not fixed and definite, but that on different days he walked, generally speaking, different numbers of miles. When, in mathematics, we speak of a "variable," what we mean is that we are considering a class of definite objects—for instance, the class of men alive at the present moment—and want to say something about *any one* of them indefinitely. Suppose that we say: "If it rains, Mr. A. will take his umbrella out with him"; the letter "A" here is what we call the sign of the "variable." We do not mean that the above proposition is about a *variable man.* There is no such thing; we say that a man varies in health and so *in time,* but, whether or not such a phrase is strictly correct, the meaning we would have to give the phrase "a variable" in the above sentence is not one and the same man at different periods of his own existence, but one and the same man who is different men in turn. What we mean is that if "A" denotes any man, and not Smith or Jones or Robinson *alone,* then he takes out his umbrella on certain occasions. The statement is not always true; it depends on A. If "A" stands for a bank manager, the statement may be true; if for a tramp or a savage, it probably is not. Instead of "A," we may put "B" or "C" or "X"; the kind of mark on paper does not really matter in the least. But we attach, by convention, certain meanings to certain signs; and so, if we wrote down a mark of exclamation for the sign of a variable, we might be misunderstood and even suspected of trying to be funny. We shall see, in the seventh chapter, the importance of the variable in logic and mathematics.

"Laws of nature" express the dependence upon one another of two or more variables. This idea of dependence of variables is fundamental in all scientific thought, and reaches its most thorough examination in mathematics and logic under the name of "functionality." On this point we must refer back to the second chapter. The ideas of function and variable were not prominent until the time of Descartes, and names for these ideas were not introduced until much later.

The conventions of analytical geometry as to the signs of co-ordinates in different quadrants of the plane had an important influence in the transformation of trigonometry from being a mere adjunct to a practical science. In the same notation as that used at the end of the first chapter, we may conveniently call the number $\frac{AP}{OP}$, which is the same for all lengths of OP, by the name "u," for short, and define $\frac{PM}{OP}$ and $\frac{OM}{OP}$ as the "sine of u," and the "cosine of u" respectively. Thus "sin u" and "cos u," as we write them for short, stand for numerical functions of u. Considering O as the origin of a system of rectangular co-ordinates of which OA is the x-axis, so that u measures the angle POA and $\frac{x}{r}$ and $\frac{y}{r}$ are cos u and sin u respectively. Now, even if u becomes so great that POA is successively obtuse, more than two right angles . . . , these definitions can be preserved, if we pay attention to the signs of x and y in the various quadrants. Thus sin u and cos u become separated from geometry, and appear as numerical functions of the variable u, whose values, as we see on reflection, repeat themselves at regular intervals as u becomes larger and larger. Thus, suppose that OP turns about O in a direction opposite to that in which the hands of a clock move. In the first quadrant, sin u

and cos u are $\frac{y}{r}$ and $\frac{x}{r}$; in the second they are $\frac{y}{r}$ and $\frac{-x}{r}$; in the third they are $\frac{-y}{r}$ and $\frac{-x}{r}$; in the fourth they are $\frac{-y}{r}$ and $\frac{x}{r}$; in the fifth they are $\frac{y}{r}$ and $\frac{x}{r}$ again; and so on. Trigonometry was separated from geometry mainly by John Bernoulli and Euler, whom we shall mention later.

We will now turn to a different development of mathematics.

The ancient Greeks seem to have had something approaching a general method for finding areas of curvilinear figures. Indeed, infinitesimal methods, which allow indefinitely close approximation, naturally suggest themselves. The determination of the area of any rectilinear figure can be reduced to that of a rectangle, and can thus be completely effected. But this process of finding areas—this "method of quadratures"—failed for areas or volumes bounded by curved lines or surfaces respectively. Then the following considerations were applied. When it is impossible to find the exact solution of a question, it is natural to endeavour to approach to it as nearly as possible by neglecting quantities which embarrass the combinations, if it be foreseen that these quantities which have been neglected cannot, by reason of their small value, produce more than a trifling error in the result of the calculation. For example, as some properties of curves with respect to areas are with difficulty discovered, it is natural to consider the curves as polygons of a great number of sides. If a regular polygon be supposed to be inscribed in a circle, it is evident that these two figures, although always different, are nevertheless more and more alike according as the number of the sides of the polygon increases. Their perimeters, their areas, the solids formed by their revolving round a given axis, the angles formed by these lines, and so on, are, if not respectively equal, at any rate so much the nearer approaching to equality as the number of sides becomes increased. Whence, by supposing the number of these sides very great, it will be possible, without any perceptible error, to assign to the circumscribed circle the properties that have been found belonging to the inscribed polygon. Thus, if it is proposed to find the area of a given circle, let us suppose this curve to be a regular polygon of a great number of sides: the area of any regular polygon whatever is equal to the product of its perimeter into the half of the perpendicular drawn from the centre upon one of its sides; hence, the circle being considered as a polygon of a great number of sides, its area ought to equal the product of the circumference into half the radius. Now, this result is exactly true. However, the Greeks, with their taste for strictly correct reasoning, could not allow themselves to consider curves as polygons of an "infinity" of sides. They were also influenced by the arguments of Zeno, and thus regarded the use of "infinitesimals" with suspicion.

Zeno showed that we meet difficulties if we hold that time and space are infinitely divisible. Of the arguments which he invented to show this, the best known is the puzzle of Achilles and the Tortoise. Zeno argued that, if Achilles ran ten times as fast as a tortoise, yet, if the tortoise has (say) 1000 yards start, it could never be overtaken. For, when Achilles had gone the 1000 yards, the tortoise would still be 100 yards in front of him; by the time he had covered these 100 yards, it would still be 10 yards in front of him; and so on for ever; thus Achilles would get nearer and nearer to the tortoise, but never overtake it. Zeno invented some other subtle puzzles for much the same purpose, and they could only be discussed really satisfactorily by quite modern mathematics.

To avoid the use of infinitesimals, Eudoxus (408–355 B.C.) devised a method, exposed by Euclid in the Twelth Book of his *Elements* and used by Archimedes to demonstrate many of his great discoveries, of verifying results found by the doubtful infinitesimal considerations. When the Greeks wished to discover the area bounded by a curve, they regarded the curve as the fixed boundary to which the inscribed and circumscribed polygons approach continually, and as much as they pleased, according as they increased the number of their sides. Thus they exhausted in some measure the space comprised between these polygons and the curve, and doubtless this gave to this operation the name of "the method of exhaustion." As these polygons terminated by straight lines were known figures, their continual approach to the curve gave an idea of it more and more precise, and "the law of continuity" serving as a guide, the Greeks could eventually arrive at the exact knowledge of its properties. But it was not sufficient for geometricians to have observed, and, as it were, guessed at these properties; it was necessary to verify them in an unexceptionable way; and this they did by proving that every supposition contrary to the existence of these properties would necessarily lead to some contradiction: thus, after, by infinitesimal considerations, they had found the area (say) of a curvilinear figure to be *a*, they verified it by proving that, if it is not *a*, it would yet be greater than the area of some polygon inscribed in the curvilinear figure whose area is palpably greater than that of the polygon.

In the seventeenth century, we have a complete contrast with the Grecian spirit. The method of discovery seemed much more important than correctness of demonstration. About the same time as the invention of analytical geometry by Descartes came the invention of a method for finding the areas of surfaces, the positions of the centres of gravity of variously shaped surfaces, and so on. In a book published in 1635, and in certain later works, Bonaventura Cavalieri (1598–1647) gave his "method of indivisibles" in which the cruder ideas of his predecessors, notably of Kepler (1571–1630), were developed. According to Cavalieri, a line is made up of an infinite number of points, each without magnitude, a surface of an infinite number of lines, each without breadth, and a volume of an infinite number of surfaces, each without thickness. The use of this idea may be illustrated by a single example. Suppose it is required to find the area of a right-angled triangle. Let the base be made up of n points (or indivisibles), and similarly let the side perpendicular to the base be made of na points, then the ordinates at the successive points of the base will contain a, $2a$ . . . , na points. Therefore the number of points in the areas is $a + 2a + \ldots + na$; the sum of which is $\frac{1}{2}(n^2a + na)$. Since n is very large, we may neglect $\frac{1}{2}na$, for it is inconsiderable compared with $\frac{1}{2}n^2a$. Hence the area is composed of a number $\frac{1}{2}(na)n$ of points, and thus the area is measured in square units by multiplying half the linear measure of the altitude by that of the base. The conclusion, we know from other facts, is exactly true.

Cavalieri found by this method many areas and volumes and the centres of gravity of many curvilinear figures. It is to be noticed that both Cavalieri and his successors held quite clearly that such a supposition that lines were composed of points was literally absurd, but could be used as a basis for a direct and concise method of abbreviation which replaced with advantage the indirect, tedious, and rigorous methods of the ancient Greeks. The logical difficulties in the principles of this and allied methods were strongly felt and commented on by philosophers—sometimes with intelligence; felt and boldly overcome by mathematicians in their strong and not unreasonable faith; and only satisfactorily solved by mathematicians—not the philosophers—in comparatively modern times.

The method of indivisibles—whose use will be shown in the next chapter in an important question of mechanics—is the same in principle as "the integral calculus." The integral calculus grew out of the work of Cavalieri and his successors, among whom the greatest are Roberval (1602–1675), Blaise Pascal (1623–1662), and John Wallis (1616–1703), and mainly consists in the provision of a convenient and suggestive notation for this method. The discovery of the infinitesimal calculus was completed by the discovery that the inverse of the problem of finding the areas of figures enclosed by curves was the problem of drawing tangents to these curves, and the provision of a convenient and suggestive notation for this inverse and simpler method, which was, for certain historical reasons, called "the differential calculus."

Both analytical geometry and the infinitesimal calculus are enormously powerful instruments for solving geometrical and physical problems. The secret of their power is that long and complicated reasonings can be written down and used to solve problems almost mechanically. It is the merest superficiality to despise mathematicians for busying themselves, sometimes even consciously, with the problem of economising thought. The powers of even the most god-like intelligences amongst us are extremely limited, and none of us could get very far in discovering any part whatever of the Truth if we could not make trains of reasoning which we have thought through and verified, very ready for and easy in future application by being made as nearly mechanical as possible. In both analytical geometry and the infinitesimal calculus, all the essential properties of very many of the objects dealt with in mathematics, and the essential features of very many of the methods which had previously been devised for dealing with them are, so to speak, packed away in a well-arranged (and therefore readily got at) form, and in an easily usable way.

CHAPTER IV

THE BEGINNINGS OF THE APPLICATION OF MATHEMATICS TO NATURAL SCIENCE—THE SCIENCE OF DYNAMICS

THE end of very much mathematics—and of the work of many eminent men—is *the simple and, as far as may be, accurate description* of things in the world around us, of which we become conscious through our senses.

Among these things, let us consider, say, a particular person's face, and a billiard ball. The appearance to the eye of the ball is obviously much easier to describe than that of the face. We can call up the image—a very accurate one—of a billiard ball in the mind of a person who has never seen it by merely giving the colour and radius. And, unless we are engaged in microscopical investigations, this description is usually enough. The description of a face is a harder matter: unless we are skilful modellers, we cannot do this even approximately; and even a good picture does not attempt literal accuracy, but only conveys a correct impression—often better than a model, say in wax, does.

Our ideal in natural science is to build up a working model of the universe out of the sort of ideas that all people carry about with them everywhere "in their heads," as we say, and to which ideas we appeal when we try to teach mathematics. These ideas are those of *number, order,* the numerical measures of *times* and *distances,* and so on. One reason why we strive after this ideal is a very practical one. If we have a working model of, say, the solar system, we can tell, in a few minutes, what our position with respect to the other planets will be at all sorts of far future times, and can thus *predict certain future events.* Everybody can see how useful this is: perhaps those persons who see it most clearly are those sailors who use the *Nautical Almanac.* We cannot make the earth tarry in its revolution round its axis in order to give us a longer day for finishing some important piece of work; but, by finding out the unchanging laws concealed in the phenomena of the motions of earth, sun, and stars, the mathematician can construct the model just spoken of. And the mathematician is completely master of his model; he can repeat the occurrences in his universe as often as he likes; something like Joshua, he can make his "sun" stand still, or hasten, in order that he may publish the *Nautical Almanac* several years ahead of time. Indeed, the "world" with which we have to deal in theoretical or mathematical mechanics is but a mathematical scheme, the function of which it is to imitate, by logical consequences of the properties assigned to it by definition, certain processes of nature as closely as possible. Thus our "dynamical world" may be called a model of reality, and must not be confused with the reality itself.

That this model of reality is constructed solely out of logical conceptions will result from our conclusion that mathematics is based on logic, and on logic alone; that such a model is possible is really surprising on reflection. The need for completing facts of nature in thought was, no doubt, first felt as a *practical* need—the need that arises because we feel it convenient to be able to predict certain kinds of future events. Thus, with a purely mathematical model of the solar system, we can tell, with an approximation which depends upon the completeness of the model, the relative positions of the sun, stars, and planets several years ahead of time; this it is that enables us to publish the *Nautical Almanac*, and makes up to us, in some degree, for our inability "to grasp this sorry scheme of things entire . . . and remould it nearer to the heart's desire."

Now, what is called "mechanics" deals with a very important part of the structure of this model. We spoke of a billiard ball just now. Everybody gets into the way, at an early age, of abstracting from the colour, roughness, and so on, of the ball, and forming for himself the conception of a *sphere*. A sphere can be exactly described; and so can what we call a "square," a "circle" and an "ellipse," in terms of certain conceptions such as those called "point," "distance," "straight line," and so on. Not so easily describable are certain other things, like a person or an emotion. In the world of moving and what we roughly class as *inanimate* objects—that is to say, objects whose behaviour is not perceptibly complicated by the phenomena of what we call "life" and "will"—people have sought from very ancient times, and with increasing success, to discover rules for the motions and rest of given systems of objects (such as a lever or a wedge) under given circumstances (pulls, pressures, and so on). Now, this discovery means: the discovery of an ideal, exactly describable motion which should approximate as nearly as possible to a natural motion or class of motions. Thus Galileo (1564–1642) discovered the approximate law of bodies falling freely, or on an inclined plane, near the earth's surface; and Newton (1642–1727) the still more accurate law of the motions of any number of bodies under any forces.

Let us now try to think clearly of what we mean by such a rule, or, as it is usually called, a "scientific" or "natural law," and why it plays an important part in the arrangement of our knowledge in such a convenient way that we can at once, so to speak, lay our hand on any particular fact the need of which is shown by practical or theoretical circumstances.

For this purpose, we will see how Galileo, in a work published in 1638, attacked the problem of a falling body. Consider a body falling freely to the earth: Galileo tried to find out, not *why* it fell, but *how* it fell—that is to say, in what mathematical form the distance fallen through and the velocity attained depends on the time taken in falling and the space fallen through. Freely falling bodies are followed with more difficulty by the eye the farther they have fallen; their impact on the hand receiving them is, in like measure, sharper; the sound of their striking louder. The velocity accordingly increases with the time elapsed and the space traversed. Thus, the modern inquirer would ask: What function is the number (v) representing the velocity of those (s and t) representing the distance fallen through and the time of falling? Galileo asked, in his primitive way: Is v proportional to s; or again, is v proportional to t? Thus he made assumptions, and then *ascertained by actual trial the correctness or otherwise of these assumptions.*

One of Galileo's assumptions was, thus, that the velocity acquired in the descent is proportional to the time of the descent. That is to say, if a body falls once, and then falls again during twice as long an interval of time as it first fell, it will attain in the second instance double the velocity it acquired in the first. To find by experiment whether or not this assumption accorded with observed facts, as it was difficult to prove by any direct means that the velocity acquired was proportional to the time of descent, but easier to investigate by what law the distance increased with the time, *Galileo deduced from his assumption the relation that obtained between the distance and the time.* This very important deduction he effected as follows.

On the straight line OA, let the abscissæ OE, OC, OG, and so on, represent in length various lengths of time elapsed from a certain instant represented by O, and let the ordinates EF, CD, GH, and so on, corresponding to these abscissæ, represent in length the magnitude of the velocities acquired at the time represented by the respective abscissæ.

We observe now that, by our assumption, *O, F, D, H, lie in a straight line OB*, and so: (1) At the instant C, at which one-half OC of the time of descent OA has elapsed, the velocity CD is also one-half of the final velocity AB; (2) If E and G are equally distant in opposite directions on OA from C, the velocity GH exceeds the mean velocity CD by the same amount that the velocity EF falls short of it; and for every instant antecedent to C there exists a corresponding one subsequent to C and equally distant from it. Whatever loss, therefore, as compared with uniform motion with half the final velocity, is suffered in the first half of the motion, such loss is made up in the second half. The distance fallen through we may consequently regard as having been uniformly described with half the final velocity.

In symbols, if the number of units of velocity acquired in t units of time is v, and suppose that v is proportional to t, the number s of units of space descended through is proportional to $\frac{1}{2}t^2$. In fact, s is given by $\frac{1}{2}vt$, and, as v is proportional to t, s is proportional to $\frac{1}{2}t^2$.

Now, Galileo verified this relation between s and t experimentally. The motion of free falling was too quick for Galileo to observe accurately with the very imperfect means—such as water-clocks—at his disposal. There were no mechanical clocks at the beginning of the seventeenth century; they were first made possible by the dynamical knowledge of which Galileo laid the foundations. Galileo, then, made the motion slower, so that s and t were big enough to be measured by rather primitive apparatus in which the moving balls ran down grooves in inclined planes. That the spaces traversed by the ball are proportional to the squares of the measures of the times in free descent as well as in motion on an inclined plane, Galileo verified by experimentally proving that a ball which falls through the height of an inclined plane attains the same final velocity as a ball which falls through its length. This experiment was an ingenious one with a pendulum whose string, when half the swing had been accomplished, caught on a fixed nail so placed that the remaining half of the swing was with a shorter string than the other half. This experiment showed that the bob of the pendulum rose, in virtue of the velocity acquired in its descent, *just as high* as it had fallen. This fact is in agreement with our instinctive knowledge of natural events; for if a ball which falls down the length of an inclined plane could attain a greater velocity than one which falls through its height, we should only have to let the body pass with the acquired velocity to another more inclined plane to make it rise to a greater vertical height than that from which it had fallen. Hence we can deduce, from the acceleration on an inclined plane, the acceleration of free descent, for, since the final velocities are the same and $s = \frac{1}{2}vt$, the lengths of the sides of the inclined plane are simply proportional to the times taken by the ball to pass over them.

The motion of falling that Galileo found actually to exist is, accordingly, a motion of which the velocity increases proportionally to the time.

Like Galileo, we have started with the notions familiar to us (through the practical arts, for example), such as that of *velocity*. Let us consider this motion more closely.

If a motion is *uniform* and c feet are travelled over in every second, at the end of t seconds it will have travelled ct feet. Put $ct = s$ for short. Then we call the "velocity" of the moving body the distance traversed in a unit of time so that it is $\frac{s}{t}$ units of length per second, the number which is the measure of the distance divided by the number which is the measure of the time elapsed. Galileo, now, attained to the conception of a motion in which the velocity increases proportionally to the time. If we draw a diagram and set off, from the origin O along the x-axis OA, a series of abscissæ which represent the times in length, and erect the corresponding ordinates to represent the velocities, the ends of these ordinates will lie on a line OB, which, in the case of the "uniformly accelerated motion" to which Galileo attained, is *straight*, as we have already seen. But if the ordinates represent *spaces* instead of *velocities*, the straight line OB becomes a curve. We see the distinction between the "curve of spaces" and "the curve of velocities," with times as abscissæ in both cases. If the velocity is uniform, the curve of spaces is a straight line OB drawn from the origin O, and the curve of velocities is a straight line parallel to the x-axis. If the velocity is variable, the curve of spaces is never a straight line; but if the motion is uniformly accelerated, the curve of velocities is a straight line like OB. The relations between the curve of spaces, the curve of velocities, and the areas of such curves AOB are, as we shall see, relations which are at once expressible by the "differential and integral calculus"—indeed, it is mainly because of this important illustration of the calculus that the elementary problems of dynamics have been treated here. And the measurement of velocity in the case where the velocity varies from time to time is an illustration of the formation of the fundamental conception of the differential calculus.

It may be remarked that the finding of the velocity of a particle at a given instant and the finding of a tangent to a curve at a given point are both of them the same kind of problem—the finding of the "differential quotient" of a function. We will now enter into the matter more in detail.

Consider a curve of spaces. If the motion is uniform, the number measuring *any* increment of the distance divided by the number measuring the corresponding increment of the time gives the same value for the measure of the velocity. But if we were to proceed like this where the velocity is variable, we should obtain widely differing values for the velocity. However, the smaller the increment of the time, the more nearly does the bit of the curve of spaces which corresponds to this increment approach straightness, and hence uniformity of increase (or decrease) of s. Thus, if we denote the increment of t by "Δt,"—where "Δ" does not stand for a number but for the phrase "the increment of,"—and the corresponding increment (or decrement) of s by "Δs," we may define the measure of average velocity in this element of the motion as $\frac{\Delta s}{\Delta t}$. But, however small Δt is, the line represented by Δs is not, usually at least, quite straight, and the velocity at the instant t, which, in the language of Leibniz's differential calculus, is defined as the quotient of "infinitely small" increments and symbolised by $\frac{ds}{dt}$, —the Δ's being replaced by d's when we consider "infinitesimals,"—appears to be only defined approximately. We have met this difficulty when considering the method of indivisibles, and will meet it again when considering the infinitesimal calculus, and will only see how it is overcome when we have become familiar with the conception of a "limit."

This new notion of velocity includes that of uniform velocity as a particular case. In fact, the rules of the infinitesimal calculus allow us to conclude, from the equation $\frac{ds}{dt} = a$, where a is some constant, the equation $s = at + b$, where b is another constant. We must remember that all this was not *expressly* formulated until about fifty years after Galileo had published his investigations on the motion of falling.

If we consider the curve of velocities, uniformly accelerated motion occupies in it exactly the same place as uniform velocity does in the curve of spaces. If we denote by v the numerical measure of the velocity at the end of t units of time, the acceleration, in the notation of the differential calculus, is measured by $\frac{dv}{dt}$, and the equation $\frac{dv}{dt} = h$, where h is some constant, is the equation of uniformly accelerated motion. In Newtonian dynamics, we have to consider *variably* accelerated motions, and this is where the infinitesimal calculus or some practically equivalent calculus such as Newton's "method of fluxions" becomes so necessary in theoretical mechanics.

We will now consider the curve of spaces for uniformly accelerated motion. On this diagram—the arcs being t and s—we will draw the curve

$$s = \frac{gt^2}{2},$$

where g denotes a constant. Of course, this is the same thing as drawing the curve $y = \frac{gx^2}{2}$ in a plane divided up by the x-axis and the y-axis of

Descartes. This curve is a parabola passing through the origin. An interesting thing about this curve is that it is the curve that would be described by a body projected obliquely near the surface of the earth if the air did not resist, and is very nearly the path of such a projectile in the resisting atmosphere. A free body, according to Galileo's view, always falls towards the earth with a uniform vertical acceleration measured by the above number g. If we project a body vertically upwards with the initial velocity of c units, its velocity at the end of t units of time is $-c+gt$ units, for if the direction downwards (of g) is reckoned positive, the direction upwards (of c) must be reckoned negative. If we project a body horizontally with the velocity of a units, and neglect the resistance of the air, Galileo recognised that it would describe, in the horizontal direction, a distance of at units in t units of time, while *simultaneously* it would fall a distance of $\frac{gt^2}{2}$ units. The two motions are to be considered as going on *independently* of each other. Thus also, oblique projection may be considered as compounded of a horizontal and a vertical projection. In all these cases the path of the projectile is a parabola; in the case of the horizontal projection, its equation in x and y co-ordinates is got from the two equations $x = at$ and $y = \frac{gt^2}{2}$, and is thus $y = \frac{gx^2}{2a^2}$.

Now, suppose that the velocity is neither uniform nor increases uniformly, but is different and increases at a different rate at different points of time. Then in the curve of velocities, the line OB is no longer straight. *In the former case, the number s was the number of square units in the area of the triangle AOB.* In this case the figure AOB is not a triangle, though we shall find that its area is the s units we seek, although v does not increase uniformly from O to A.

Notice again that if, on OA, we take points C and E very close together, the little arc DF is very nearly straight, and the figure DGF very nearly a rectilinear triangle. Note that we are only trying, in this, to get a first approximation to the value of s, so that, instead of the continuously changing velocities we know—or think we know—from our daily experience, we are considering a fictitious motion in which the velocity increases (or decreases) so as to be the same as that of the motion thought of at a large number of points at minute and equal distances, and between successive points increases (or decreases) uniformly.

Note also that we are assuming (what usually happens with the curves with which we shall have to do) that the arc DF which corresponds to CE becomes as straight as we wish if we take C and E close enough together.

And now let us calculate s approximately. Starting from O, in the first small interval OH the rectilinear triangle OHK, where HK is the ordinate at H, represents approximately the space described. In the next small interval HL, where the length of HL is equal to that of OH, the space described is represented by the rectilinear figure $KHLM$. The rectangle KL is the space passed over with the uniform velocity HK in time HL; and the triangle KNM is the space passed over by a motion in which the velocity increases from zero to MN. And so on for other intervals beyond HL. Thus s is ultimately given (approximately) as the number of square units in a polygon which closely approximates to the figure AOB.

We must now say a few words about the meaning of the letters in geometrical and mechanical *equations* which, following Descartes, we use instead of the proportions used by Galileo and even many of his contemporaries and followers. It seems better, when beginning mechanics, to think in proportions, but afterwards, for convenience in dealing with the symbolism of mathematical data, it is better to think in equations.

A typical proportion is: Final velocities are to one another as the times; or, in symbols,

$$\text{“}V : V' : : T : T'.\text{”}$$

Here "V" (for example) is just short for "the velocity attained at the end of the period of time" (reckoned from some fixed instant) denoted by "T," and $V : V'$, and $T : T'$, are just *numbers* (real numbers); and the proportion states the equality of these numbers. Hence the proportion is sometimes written "$V : V' = T : T'$." If, now, v is the numerical measure, merely, of V, v' that of V', and so on, we have $\frac{v}{v'} = \frac{t}{t'}$ or $vt' = v't$.

In the last equation, the letters v and t have a mnemonic significance, as reminding us that we started from *velocities* and *times, but we must carefully avoid the idea that we are "multiplying"* (or can do so) *velocities by times;* what we *are* doing is multiplying the numerical measures of them. People who write on geometry and mechanics often say inaccurately, simply for shortness, "Let s denote the distance, t the time," and so on; whereas, by a tacit convention, small italics are usually employed to denote *numbers*. However, in future, for the sake of shortness, I shall do as the writers referred to, and speak of v as "the velocity." Equations in mechanics, such as "$s = \frac{gt^2}{2}$." are only possible if the left-hand side is of the same kind as the right-hand side: we cannot equate spaces and times, for example.

Suppose that we have fixed on the unit of length as one inch and the unit of time as one second. As unit of velocity we might choose the velocity with which, say, a inches are described uniformly in one second. If we did this, we should express the relation between the s units of space passed over by a body with a given velocity (v units) in a given time (t units) as "$s = avt$"; whereas, if we defined the unit of velocity as the velocity with which the unit of length is travelled over in the unit of time, we should write "$s = vt$."

Among the units derived from the fundamental units—such as those of length and time—the simplest possible relations are made to hold. Thus, as the unit of area and the unit of volume, the square and the cube of unit sides are respectively used, the unit of velocity is the uniform rate at which unit of length is travelled over in the unit of time, the unit of acceleration is the gain of unit velocity in unit time, and so on.

The derived units depend on the fundamental units, and the *function* which a given derived unit is of its fundamental units is called its "dimensions." Thus the velocity v is got by dividing the length s by the time t. The dimensions of a velocity are written

$$\text{``}[V] = \frac{[L]}{[T]},\text{''}$$

and those of an acceleration—denoted "F"—

$$\text{``}[F] = \frac{[V]}{[T]} = \frac{[L]}{[T]^2}.\text{''}$$

These equations are merely mnemonic; the letters do not mean numbers. The mnemonic character comes out when we wish to pass from one set of units to another. Thus, if we pass to a unit of length b times greater and one of time c times greater, the acceleration f with the old units is related to that (f') with the new units by the equation

$$f'\left(\frac{c^2}{b}\right) = f.$$

As the units become greater, f' becomes less; and, since the dimensions of F are $\frac{[L]}{[T]^2}$, the factor $\frac{c^2}{b}$ is obviously suggested to us—the symbol "$[T]^2$" suggesting a squaring of the number measuring the time.

From Galileo's work resulted the conclusion that, where there is no change of *velocity in a straight line,* there is no force. The state of a body unacted upon by force is uniform rectilinear motion; and rest in a special case of this motion where the velocity is and remains zero. This "law of inertia" was exactly opposite to the opinion, derived from Aristotle, that force is requisite to keep up a uniform motion, and may be roughly verified by noticing the behaviour of a body projected with a given velocity and moving under little resistance—as a stone moving on a sheet of ice. Newton and his contemporaries saw how important this law was in the explanation of the motion of a planet—say, about the sun. Think of a simple case, and imagine the orbit to be a circle. The planet tends to move along the tangent with uniform velocity, but the attraction of the sun simultaneously draws the planet towards itself, and the result of this continual combination of two motions is the circular orbit. Newton succeeded in calculating the shapes of the orbits for different laws of attraction, and found that, when attraction varies inversely as the square of the distance, the shapes are conic sections, as had been observed in the case of our solar system.

The problem of the solar system appeared, then, in a mathematical dress; various things move about in space, and this motion is completely described if we know the geometrical relations—distances, positions, and angular distances—between these things at some moment, the velocities at this moment, and the accelerations at *every* moment. Of course, if we knew all the positions of all the things at all the instants, our description would be complete; it happens that the *accelerations* are usually simpler to find directly than the positions: thus, in Galileo's case the acceleration was simply constant. Thus, we are given functional relations between these positions and their rates of change. We have to determine the positions from these relations.

It is the business of the "method of fluxions" or the "infinitesimal calculus" to give methods for finding the relations between variables from relations between their rates of change or between them and these rates. This shows the importance of the calculus in such physical questions.

Mathematical physics grew up—perhaps too much so—on the model of theoretical astronomy, its first really extensive conquest. There are signs that mathematical physics is freeing itself from its traditions, but we need not go further into the subject in this place.

Roberval devised a method of tangents which is based on Galileo's conception of the composition of motions. The tangent is the direction of the resultant motion of a point describing the curve. Newton's method, which is to be dealt with in the fifth chapter, is analogous to this, and the idea of velocity is fundamental in his "method of fluxions."

CHAPTER V

THE RISE OF MODERN MATHEMATICS —THE INFINITESIMAL CALCULUS

IN the third chapter we have seen that the ancient Greeks were sometimes occupied with the theoretically exact determination of the areas enclosed by curvilinear figures, and that they used the "method of exhaustion," and, to demonstrate the results which they got, an indirect method. We have seen, too, a "method of indivisibles," which was direct and seemed to gain in brevity and efficiency from a certain lack of correctness in expression and perhaps even a small inexactness in thought. We shall find the same merits and demerits—both, especially the merits, intensified —in the "infinitesimal calculus."

By the side of researches on quadratures and the finding of volumes and centres of gravity developed the methods of drawing tangents to curves. We have begun to deal with this subject in the third chapter: here we shall illustrate the considerations of Fermat (1601–1665) and Barrow (1630–1677)—the intellectual descendants of Kepler—by a simple example.

Let it be proposed to draw a tangent at a given point P in the circumference of a circle of centre O and equation $x^2 + y^2 = 1$. Let us take the circle to be a polygon of a great number of sides; let PQ be one of these sides, and produce it to meet the x-axis at T. Then PT will be the tangent in question. Let the co-ordinates of P be X and Y; those of Q will be $X + e$ and $Y + a$, where e and a are infinitely small increments, positive or negative. From a figure in which the ordinates and abscissæ of P and Q are drawn, so that the ordinate of P is PR, we can see, by a well-known property of triangles, that TR is to RP (or Y) as e is to a. Now, X and Y are related by the equation $X^2 + Y^2 = 1$, and, since Q is also on the locus $x^2 + y^2 = 1$, we have $(X + e)^2 + (Y + a)^2 = 1$. From the two equations in which X and Y occur, we conclude that $2eX + e^2 + 2aY + a^2 = 0$, and hence $\frac{e}{a}(X + \frac{e}{2}) + Y + \frac{a}{2} = 0$. But $\frac{e}{a} = \frac{TR}{Y}$; hence $TR = \frac{-Y(Y + {}^{a}/_{2})}{X + {}^{e}/_{2}}$.

Now, a and e may be neglected in comparison with X and Y, and thus we can say that, at any rate *very* nearly, we have $TR = \frac{Y^2}{X}$. But this is *exactly right,* for, since TP is at right angles to OP, we know that OR is to RP as PR is to RT. Here X and Y are constant, but we can say that the abscissa of the point where the tangent at *any* point (say y) of the circle cuts the x-axis is given by adding $\frac{y^2}{x}$ to x.

Thus, we can find tangents by considering the ratios of infinitesimals to one another. The method obviously applies to other curves besides circles; and Barrow's method and nomenclature leads us straight to the notation and nomenclature of Leibniz. Barrow called the triangle PQS, where S is where a parallel to the x-axis through Q meets PR, the "differential triangle," and Leibniz denoted Barrow's a and e by dy and dx (short for the "differential of y" and "the differential of x," so that "d" does not denote a number but "dx" altogether stands for an "infinitesimal") respectively, and called the collection of rules for working with his signs the "differential calculus."

But before the notation of the differential calculus and the rules of it were discovered by Gottfried Wilhelm von Leibniz (1646–1716), the celebrated German philosopher, statesman, and mathematician, he had invented the notation and found some of the rules of the "integral calculus": thus, he had used the now well-known sign "$\int$" or long "s" as short for "the sum of," when considering the sum of an infinity of infinitesimal elements as we do in the method of indivisibles. Suppose that we propose to determine the area included between a certain curve $y = f(x)$, the x-axis, and two fixed ordinates whose equations are $x = a$ and $x = b$; then, if we make use of the idea and notation of differentials, we notice that the area in question can be written as

$$\text{“}\int y \, . \, dx,\text{”}$$

the summation extending from $x = a$ to $x = b$. We will not here further concern ourselves about these boundaries. Notice that in the above expression we have put a dot between the "y" and the "dx": this is to indicate that y is to multiply dx. Hitherto we have used juxtaposition to denote multiplication, but here d is written close to x with another end in view; and it is desirable to emphasise the difference between "d" used in the sense of an adjective and "d" used in the sense of a multiplying number, at least until the student can easily tell the difference by the context. If, then, we imagine the abscissa divided into equal infinitesimal parts, each of length dx, corresponding to the constituents called "points" in the method of indivisibles, $y \, . \, dx$ is the area of the little rectangle of sides dx and y which stand at the end of the abscissa x. If, now, instead of extending to $x = b$, the summation extends to the ordinate at the indeterminate or "variable" point x, $y \, . \, dx$ becomes a function of x.

Now, if we think what must be the differential of this sum—the infinitesimal increment that it gets when the abscissa of length x, which is one of the boundaries, is increased by dx—we see that it must be $y \, . \, dx$. Hence

$$d(\int y \, . \, dx) = y \, . \, dx,$$

and hence the sign of "d" destroys, so to speak, the effect of the sign "$\int$". We also have $\int dx = x$, and find that this summation is the inverse process to differentiation. *Thus the problems of tangents and quadratures are inverses of one another.* This vital discovery seems to have been first made by Barrow without the help of any technical symbolism. The quantity which by its differentiation produces a proposed differential, is called the "integral" of this differential; since we consider it as having been formed by infinitely small continual additions: each of these additions is what we have named the differential of the increasing quantity, it is a fraction of it: and the sum of all these fractions is the entire quantity which we are in search of. For the same reason we call "integrating" or "taking the sum of" a differential the finding the integral of the sum of all the infinitely small successive additions which form the series, the differential of which, properly speaking, is the general term.

It is evident that two variables which constantly remain equal increase the one as much as the other during the same time, and that consequently their differences are equal: and the same holds good even if these two quantities had differed by any quantity whatever when they began to vary; provided that this primitive difference be always the same, their differentials will always be equal.

Reciprocally, it is clear that two variables which receive at each instant infinitely small equal additions must also either remain constantly equal to one another, or always differ by the same quantity—that is, the integrals of two differentials which are equal can only differ from each other by a constant quantity. For the same reason, if any two quantities whatever differ in an infinitely small degree from each other, their differentials will also differ from one another infinitely little: and reciprocally if two differential quantities differ infinitely little from one another, their integrals, putting aside the constant, can also differ but infinitely little one from the other.

Now, some of the rules for differentiation are as follows. If $y = f(x)$, $dy = f(x + dx) - f(x)$, in which higher powers of differentials added to lower ones may be neglected. Thus, if $y = x^2$, then $dy = (x + dx)^2 - x^2 = 2x \,.\, dx + (dx)^2 = 2x \,.\, dx$. Here it is well to refer back to the treatment of the problem of tangents at the beginning of this chapter. Again, if $y = a \,.\, x$, where a is constant, $dy = a \,.\, dx$. If $y = x \,.\, z$, then

$dy = (x + dx)(z + dz) - x \,.\, z = x \,.\, dz + z \,.\, dx$. If $y = \dfrac{x}{z}$, $x = y \,.\, z$, so

$dx = y \,.\, dz + z \,.\, dy$; hence $dy = \dfrac{dx - y \,.\, dz}{z}$. Since the integral calculus

is the inverse of the differential calculus, we have at once

$$\int 2x \,.\, dx = x^2, \int a \,.\, dx = a\int dx,$$
$$\int x \,.\, dz + \int z \,.\, dx = xz,$$

and so on. More fully, from $d(x^3) = 3x^2 \,.\, dx$, we conclude, not that $\int x^2 \,.\, dx = \frac{1}{3}x^3$, but that $\int x^2 \,.\, dx = \frac{1}{3}x^3 + c$, where "$c$" denotes some constant depending on the fixed value for x from which the integration starts.

Consider a parabola $y^2 = ax$; then $2y \,.\, dy = a \,.\, dx$, or $dx = \dfrac{2y \,.\, dy}{a}$.

Thus the area from the origin to the point x is $\displaystyle\int \frac{2y^2 \,.\, dy}{a} + c$; but $d\dfrac{2y^3}{3a} = \dfrac{2y^2 \,.\, dy}{a}$; thus the area is $\dfrac{2y^3}{3a} + c$, or, since $y^2 = ax$, $\frac{2}{3}x \,.\, y + c$. To determine c when we measure the area from 0 to x, we have the area zero when $x = 0$; hence the above equation gives $c = 0$. This whole result, now quite simple to us, is one of the greatest discoveries of Archimedes.

Let us now make a few short reflections on the infinitesimal calculus. First, the extraordinary power of it in dealing with complicated questions lies in that the question is split up into an infinity of *simpler* ones which can all be dealt with at once, thanks to the wonderfully economical fashion in which the calculus, like analytical geometry, deals with variables. Thus, a *curvilinear* area is split up into *rectangular* elements, all the rectangles are added together at once when it is observed that integral is the inverse of the easily acquired practice of differentiation. We must never lose sight of the fact that, when we differentiate y or integrate $y \,.\, dx$, we are considering, not a particular x or y, but *any* one of an infinity of them. Secondly, we have seen that what in the first place had been regarded but as a simple method of approximation, leads at any rate in certain cases to results perfectly exact. The fact is that the exact results are due to a compensation of errors: the error resulting from the false supposition made, for example, by regarding a curve as a polygon with an infinite number of sides each infinitely small and which when produced is a tangent of the curve, is corrected or compensated for by that which springs from the very processes of the calculus, according to which we retain in differentiation infinitely small quantities of the same order alone. In fact, after having introduced these quantities into the calculation to facilitate the expression of the conditions of the problem, and after having regarded them as absolutely zero in comparison with the proposed quantities, with a view to simplify these equations, in order to banish the errors that they had occasioned, and to obtain a result perfectly exact, there remains but to eliminate these same quantities from the equations where they may still be.

But all this cannot be regarded as a strict proof. There *are* great difficulties in trying to determine what infinitesimals are: at one time they are treated like finite numbers and at another like zeros or as "ghosts of departed quantities," as Bishop Berkeley, the philosopher, called them.

Another difficulty is given by differentials "of higher orders than the first." Let us take up again the considerations of the fourth chapter. We saw that $v = \frac{ds}{dt}$, and found that s was got by integration: $s = \int v \,.\, dt$. This is now an immediate inference, since $\frac{ds}{dt} dt = ds$. Now, let us substitute for v in $\frac{dv}{dt}$. Here t is the independent variable, and all of the older mathematicians treated the elements dt as constant—the interval of the independent variable was split up into atoms, so to speak, which themselves were regarded as known, and in terms of which other differentials, ds, dx, dy, were to be determined. Thus

$$\frac{dv}{dt} = \frac{d(^{ds}/_{dt})}{dt} = \frac{^1/_{dt} \,.\, d(ds)}{dt} = \frac{d^2s}{dt^2},$$

"d^2s" being written for "$d(ds)$" and "dt^2" for "$(dt)^2$". Thus the acceleration was expressed as "the second differential of the space divided by the square of dt." If $\frac{d^2s}{dt^2}$ were constant, say, a, then $\frac{d^2s}{dt} = a \,.\, dt$; and, integrating both sides:

$$\frac{ds}{dt} = \int a \,.\, dt = a\int dt = at + b,$$

where b is a new constant. Integrating again, we have:

$$s = a\int t \,.\, dt + b\int dt = \frac{at^2}{2} + bt + c,$$

which is a more general form of Galileo's result. Many complicated problems which show how far-reaching Galileo's principles are were devised by Leibniz and his school.

Thus, the infinitesimal calculus brought about a great advance in our powers of describing nature. And this advance was mainly due to Leibniz's notation; Leibniz himself attributed all of his mathematical discoveries to his improvements in notation. Those who know something of Leibniz's work know how conscious he was of the suggestive and economical value of a good notation. And the fact that we still use and appreciate Leibniz's

"∫" and "*d*," even though our views as to the principles of the calculus are very different from those of Leibniz and his school, is perhaps the best testimony to the importance of this question of notation. This fact that Leibniz's notations have become permanent is the great reason why I have dealt with his work before the analogous and prior work of Newton.

Isaac Newton (1642–1727) undoubtedly arrived at the principles and practice of a method equivalent to the infinitesimal calculus much earlier than Leibniz, and, like Roberval, his conceptions were obtained from the dynamics of Galileo. He considered curves to be described by moving points. If we conceive a moving point as describing a curve, and the curve referred to co-ordinate axes, then the velocity of the moving point can be decomposed into two others parallel to the axes of x and y respectively; these velocities are called the "fluxions" of x and y, and the velocity of the point is the fluxion of the arc. Reciprocally the arc is the "fluent" of the velocity with which it is described. From the given equation of the curve we may seek to determine the relations between the fluxions—and this is equivalent to Leibniz's problem of differentiation;—and reciprocally we may seek the relations between the co-ordinates when we know that between their fluxions, either alone or combined with the co-ordinates themselves. This is equivalent to Leibniz's general problem of integration, and is the problem to which we saw, at the end of the fourth chapter, that theoretical astronomy reduces.

Newton denoted the fluxion of x by "$\dot{x}$," and the fluxion of the fluxion (the acceleration) of $\dot{x}$ by "$\ddot{x}$." It is obvious that this notation becomes awkward when we have to consider fluxions of higher orders; and further, Newton did not indicate by his notation the independent variable considered. Thus "$\dot{y}$" might possibly mean either $\frac{dy}{dt}$ or $\frac{dy}{dx}$. We have $\dot{x} = \frac{dx}{dt}$, $\ddot{x} = \frac{d\dot{x}}{dt} = \frac{d^2x}{dt^2}$; but a dot-notation for $\frac{d^nx}{dt^n}$ would be clumsy and inconvenient. Newton's notation for the "inverse method of fluxions" was far clumsier even, and far inferior to Leibniz's "∫".

The relations between Newton and Leibniz were at first friendly, and each communicated his discoveries to the other with a certain frankness. Later, a long and acrimonious dispute took place between Newton and Leibniz and their respective partisans. Each accused—unjustly, it seems—the other of plagiarism, and mean suspicions gave rise to meanness of conduct, and this conduct was also helped by what is sometimes called "patriotism." Thus, for considerably more than a century, British mathematicians failed to perceive the great superiority of Leibniz's notation. And thus, while the Swiss mathematicians, James Bernoulli (1654–1705), John Bernoulli (1667–1748), and Leonhard Euler (1707–1783), the French mathematicians d'Alembert (1707–1783), Clairaut (1713–1765), Lagrange (1736–1813), Laplace (1749–1827), Legendre (1752–1833), Fourier (1768–1830), and Poisson (1781–1850), and many other Continental mathematicians, were rapidly [6] extending knowledge by using the infinitesimal calculus in all branches of pure and applied mathematics, in England comparatively little progress was made. In fact, it was not until the beginning of the nineteenth century that there was formed, at Cambridge, a Society to introduce and spread the use of Leibniz's notation among British mathematicians: to establish, as it was said, "the principles of pure *d*-ism in opposition to the *dot*-age of the university."

The difficulties met and not satisfactorily solved by Newton, Leibniz, or their immediate successors, in the principles of the infinitesimal calculus, centre about the conception of a "limit"; and a great part of the meditations of modern mathematicians, such as the Frenchman Cauchy (1789–1857), the Norwegian Abel (1802–1829), and the German Weierstrass (1815–1897), not to speak of many still living, have been devoted to the putting of this conception on a sound logical basis.

We have seen that, if $y = x^2$, $\frac{dy}{dx} = 2x$. What we do in forming $\frac{dy}{dx}$ is to form $\frac{(x + \Delta x)^2 - x^2}{\Delta x}$, which is readily found to be $2x + \Delta x$, and then consider that, as Δx approaches 0 more and more, the above quotient approaches $2x$. We express this by saying that the "limit, as h [Δx] approaches 0," is $2x$. We do not consider Δx as being a fixed "infinitesimal" or as an absolute zero (which would make the above quotient become indeterminate $\frac{0}{0}$), nor need we suppose that the quotient *reaches* its limit (the state of Δx being 0). What we need to consider is that "Δx" should represent a variable which can take values differing from 0 by as little as we please. That is to say, if we choose *any* number, however small, there is a value which Δx can take, and which differs from 0 by less than that

[6] It is difficult for a mathematician not to think that the sudden and brilliant dawn on eighteenth-century France of the magnificent and apparently all-embracing physics of Newton and the wonderfully powerful mathematical method of Leibniz inspired scientific men with the belief that the goal of all knowledge was nearly reached and a new era of knowledge quickly striding towards perfection begun; and that this optimism had indirectly much to do in preparing for the French Revolution.

number. As before, when we speak of a "variable" we mean that we are considering a certain *class.* When we speak of a "limit," we are considering a certain *infinite* class. Thus the sequence of an infinity of terms 1, $\frac{1}{2}$, $\frac{1}{4}$, $\frac{1}{8}$, $\frac{1}{16}$, and so on, whose law of formation is easily seen, has the limit 0. In this case 0 is such that any number greater than it is greater than some term of the sequence, but 0 itself is not greater than any term of the sequence and is not a term of the sequence. A sequence like 1, $1 + \frac{1}{2}$, $1 + \frac{1}{2} + \frac{1}{4}$, $1 + \frac{1}{2} + \frac{1}{4} + \frac{1}{8}$ . . ., has an analogous *upper* limit 2. A function $f(x)$, as the independent variable x approaches a certain value, like $\frac{2x}{x}$ as x approaches 0, may have a value (in this case 2, though *at* 0, $\frac{2x}{x}$ is indeterminate). The question of the limits of a function in general is somewhat complicated, but the most important limit is $\frac{f(x + \Delta x) - f(x)}{\Delta x}$ as Δx approaches 0; this, if $y = f(x)$, is $\frac{dy}{dx}$.

That the infinitesimal calculus, with its rather obscure "infinitesimals"—treated like finite numbers when we write $\frac{dy}{dx}\,dx = dy$ and $\frac{1}{dy/dx} = \frac{dx}{dy}$, and then, on occasion, neglected—leads so often to correct results is a most remarkable fact, and a fact of which the true explanation only appeared when Cauchy, Gauss (1777–1855), Riemann (1826–1866), and Weierstrass had developed the theory of an extensive and much used class of functions. These functions happen to have properties which make them especially easy to be worked with, and nearly all the functions we habitually use in mathematical physics are of this class. A notable thing is that the complex numbers spoken of in the second chapter *make* this theory to a great extent.

Large tracts of mathematics have, of course, not been mentioned here. Thus, there is an elaborate theory of integer numbers to be referred to in a note to the seventh chapter, and a geometry using the conceptions of the ancient Greeks and methods of modern mathematical thought; and very many men still regard space-perception as something mathematics deals with. We will return to this soon. Again, algebra has developed and branched off; the study of functions in general and in particular has grown; and soon a list of some of the many great men who have helped in all this would not be very useful. Let us now try to resume what we have seen of the development of mathematics along what seem to be its main lines.

In the earliest times men were occupied with particular questions—the properties of particular numbers and geometrical properties of particular figures, together with simple mechanical questions. With the Greeks, a more general study of classes of geometrical figures began. But traces of an earlier exception to this study of particulars are afforded by "algebra." In it and its later form symbols—like our present x and y—took the place of numbers, so that, what is a great advance in economy of thought and other labour, a part of calculation could be done with symbols instead of numbers, so that the one result stated, in a manner analogous to that of Greek geometry, a proposition valid for a whole infinite class of different numbers.

The great revolution in mathematical thought brought about by Descartes in 1637 grew out of the application of this general algebra to geometry by the very natural thought of substituting the numbers expressing the lengths of straight lines for those lines. Thus a point in a plane—for instance—is determined in position by two numbers x and y, or co-ordinates. Now, as the point in question varies in position, x and y both vary; to every x belongs, in general, one or more y's, and we arrive at the most beautiful idea of a single algebraical equation between x and y representing the whole of a curve—the one "equation of the curve" expressing the general law by which, given any particular x out of an infinity of them, the corresponding y or y's can be found.

The problem of drawing a tangent—the limiting position of a secant, when the two meeting points approach indefinitely close to one another—at any point of a curve came into prominence as a result of Descartes' work, and this, together with the allied conceptions of velocity and acceleration "at an instant," which appeared in Galileo's classical investigation, published in 1638, of the law according to which freely falling bodies move, gave rise at length to the powerful and convenient "infinitesimal calculus" of Leibniz and the "method of fluxions" of Newton. Mathematically, the finding of the tangent at the point of a curve, and finding the velocity of a particle describing this curve when it gets to that point, are identical problems. They are expressed as finding the "differential quotient," or the "fluxion" at the point. It is now known to be very probable that the above two methods, which are theoretically—but not practically—the same, were discovered independently; Newton discovered his first, and Leibniz published his first, in 1684. The finding of the areas of curves and of the shapes of the curves which moving particles describe under given forces showed themselves, in this calculus, as results of the inverse process to that of the direct process which serves to find tangents and the law of attraction to a given point from the datum of the path described by a particle. The direct process is called "differentiation," the inverse process "integration."

Newton's fame is chiefly owing to his application of this method to the solution, which, in its broad outlines, he gave of the problem of the motion of the bodies in the solar system, which includes his discovery of the law according to which all matter gravitates towards—is attracted by—other matter. This was given in his *Principia* of 1687; and for more than a century afterwards mathematicians were occupied in extending and applying the calculus.

Then came more modern work, more and more directed towards the putting of mathematical methods on a sound logical basis, and the separation of mathematical processes from the sense-perception of space with which so much in mathematics grew and grows up. Thus trigonometry took its place by algebra as a study of certain mathematical functions, and it began to appear that the true business of geometry is to supply beautiful and suggestive pictures of abstract—"analytical" or "algebraical" or even "arithmetical," as they are called—processes of mathematics. In the next chapter we shall be concerned with part of the work of logical examination and reconstruction.

CHAPTER VI

MODERN VIEWS OF LIMITS AND NUMBERS

LET us try to form a clear idea of the conception which showed itself to be fundamental in the principles of the infinitesimal calculus, the conception of a *limit*.

Notice that the limit of a sequence is a number which is already defined. We cannot prove that there is a limit to a sequence unless the limit sought is among the numbers already defined. Thus, in the system of "numbers" —here we must refer back to the second chapter—consisting of all fractions (or ratios), we can say that the sequence (where 1 and 2 are written for the ratios $1/1$ and $2/1$) $1, 1 + 1/2, 1 + 1/2 + 1/4, \ldots$, has a limit (2), but that the sequence

$$1, 1 + 4/10, 1 + 4/10 + 1/100, 1 + 4/10 + 1/100 + 4/1000, \ldots,$$

$$\text{or } 1 \cdot 4142 \ldots,$$

got by extracting the square root of 2 by the known process of decimal arithmetic, has not. In fact, it can be proved that there is no ratio such that it is a limit for the above sequence. If there were, and it were denoted by "x," we would have $x^2 = 2$. Here we come again to the question of incommensurables and "irrational numbers." The Greeks were quite right in distinguishing so sharply between numbers and magnitudes, and it was a tacit, natural, and unjustified—not, as it happens, incorrect—presupposition that the series of numbers, completed into the series of what are called "real numbers," exactly corresponds to the series of points on a straight line. The series of points which represents the sequence last named seems undoubtedly to possess a limit; this limiting point was assumed to represent some number, and, since it could not represent an integer or a ratio, it was said to represent an "irrational number," $\sqrt{2}$. Another irrational number is that which is represented by the incommensurable ratio of the circumference of a circle to its diameter. This number is denoted by the Greek letter "π," and its value is nearly $3\cdot1416$. . . . Of course, the process of approximation by decimals never comes to an end.

The subject of limits forced itself into a very conspicuous place in the seventeenth and eighteenth centuries owing to the use of infinite series as a means of approximate calculation. I shall distinguish what I call "sequences" and "series." A sequence is a collection—finite or infinite—of numbers; a series is a finite or infinite collection of numbers *connected by addition*. Sequences and series can be made to correspond in the following way. To the sequence 1, 2, 3, 4, . . . belongs a series of which the terms are got by subtracting, in order, the terms of the sequence from the ones immediately following them, thus:

$$(2-1)+(3-2)+(4-3)+\ldots=1+1+1+\ldots;$$

and from a series the corresponding sequence can be got by making the sum of the first, the first two, the first three, . . . terms the first, second, third . . . term of the sequence respectively. Thus, to the series $1+1+1+\ldots$ corresponds the sequence 1, 2, 3, . . .

Now, if a series has only a finite number of terms, it is possible to find the sum of all the terms; but if the series is unending, we evidently cannot. But in certain cases the corresponding sequence has a limit, and this limit is called by mathematicians, neither unnaturally nor accurately, "the sum to infinity of the series." Thus, the sequence $1,\ 1+\frac{1}{2},\ 1+\frac{1}{2}+\frac{1}{4}\ \ldots$ has the limit 2, and so the sum to infinity of the series $1+\frac{1}{2}+\frac{1}{4}+\frac{1}{8}+\ldots$ is 2. Of course, all series do not have a sum: thus $1+1+1+\ldots$ to infinity has not—the terms of the corresponding sequence increase continually beyond all limits. Notice particularly that the terms of a sequence may increase continually, and yet have a limit—those of the above sequence with limit 2 so increase, but not beyond 2, though they do beyond any number less than 2; also notice that the terms of a sequence may increase beyond all limits even if the terms of the corresponding series continually diminish, remaining positive, towards 0. The series $1+\frac{1}{2}+\frac{1}{3}+\frac{1}{4}+\frac{1}{5}+\ldots$ is such a series; the terms of the sequence slowly increase beyond all limits, as we see when we reflect that the sums

$$\tfrac{1}{3}+\tfrac{1}{4},\ \tfrac{1}{5}+\tfrac{1}{6}+\tfrac{1}{7}+\tfrac{1}{8},\ \tfrac{1}{9}+\ldots+\tfrac{1}{16},\ \ldots$$

are all greater than $\frac{1}{2}$. It is very important to realise the fact illustrated by this example; for it shows that the conditions under which an infinite series has a sum are by no means as simple as they might appear at first sight.

The logical scrutiny to which, during the last century, the processes and conceptions of mathematics have been subjected, showed very plainly that it was a sheer assumption that such a process as 1·4142 . . . , though *all* its terms are less than 2, for example, has any limit at all. When we replace numbers by points on a straight line, we feel fairly sure that there is one point which behaves to the points representing the above sequence in the same sort of way as 2 to the sequence 1, 1 + ½, 1 + ½ + ¼, Now, if our system of numbers is to form a *continuum*, as a line seems to our thoughts to be; so that we can affirm that our number system is adequate, when we introduce axes in the manner of analytical geometry, to the description of all the phenomena of change of position which take place in our space,[7] *then* we must have a number $\sqrt{2}$ which is the limit of the sequence 1·4142 . . . if 2 is of the series 1 + ½ + ¼ + . . . , for to every point of a line must correspond a number which is subject to the same rules of calculation as the ratios or integers. Thus we must, to justify from a logical point of view our procedure in the great mathematical methods, show what irrationals are, and define them *before* we can prove that they are limits. We cannot take a series, whose law is evident, which has no ratio for sum, and yet such that the terms of the corresponding sequence all remain less than some fixed number (such as $1 + \frac{1}{1} + \frac{1}{1\cdot 2} + \frac{1}{1\cdot 2\cdot 3} + \frac{1}{1\cdot 2\cdot 3\cdot 4} + \ldots$, when all the terms of the corresponding sequence are less than 3, for example), and then say that it "defines a limit." All we can prove is that *if* such a series has a limit, then, if the terms of its corresponding sequence do not decrease as we read from left to right (as in the preceding example), it cannot have more than one limit.

Some mathematicians have simply *postulated* the irrationals. At the beginning of their discussions they have, tacitly or not, said: "In what follows we will *assume* that there are such things as fill up kinds of gaps in the system of rationals (or ratios)." Such a gap is shown by this. The rationals less than ½ and those greater than ½ form two sets and ½ divides them. The rationals x such that x^2 is greater than 2 and those x's such that x^2 is less than 2 form two analogous sets, but there is only an analogue to the dividing number ½ if we postulate a number $\sqrt{2}$. Thus by

[7] The only kind of change dealt with in the science of mechanics is change of position, that is, motion. It does not seem to me to be necessary to adopt the doctrine that the complete description of any physical event is of a mechanical event; for it is possible to assign and calculate with numbers of our number-continuum to other varying characteristics (such as temperature) of the state of a body besides position.

postulation we fill up these subtle gaps in the set of rationals and get a continuous set of real numbers. But we can avoid this postulation if we define "$\sqrt{2}$" as the name of the class of rationals x such that x^2 is less than 2 and "(½)" as the name of the class of rationals x such that x is less than ½. Proceeding thus, we arrive at a set of *classes*, some of which correspond to rationals, as (½) to ½, but the rest satisfy our need of a set without gaps. There is no reason why we should not say that these classes *are* the *real numbers* which include the irrationals. But we must notice that rationals are never real numbers; ½ is not (½), though analogous to it. We have much the same state of things as in the second chapter, where 2, +2 and 2/1 were distinguished and then deliberately confused because, with the mathematicians, we felt the importance of analogy in calculation. Here again we identify (½) with ½, and so on.

Thus, integers, positive and negative "numbers," ratios, and real "numbers" are all different things: real numbers are classes, ratios and positive and negative numbers are relations. Integers, as we shall see, are classes. Very possibly there is a certain arbitrariness about this, but this is unimportant compared with the fact that in modern mathematics we have reduced the definitions of all "numbers" to logical terms. Whether they are classes or relations or propositions or other logical entities is comparatively unimportant.

Integers can be defined as certain classes. Mathematicians like Weierstrass stopped before they got as far as this: they reduced the other numbers of analysis to logical developments out of the conception of integer, and thus freed analysis from any remaining trace of the sway of geometry. But it was obvious that integers had to be defined, if possible, in logical terms. It has long been recognised that two collections consist of the same number of objects if, and only if, these collections can be put in such a relation to one another that to every object of each one belongs one and only one object of the other. We must not think that this implies that we have already the idea of the *number one*. It is true that "one and only one" seems to use this idea. But "the class a has one and only one member" is simply a short way of expressing: "x is a member of a, and if y is also a member of a, then y is identical with x." It is true, also, that we use the idea of the *unity* or the *individuality* of the things considered. But this unity is a property of each individual, while the number 1 is a property of a *class*. If a class of pages of a book is itself, under the name of a "volume," a member of a class of books, the same class of pages has both a number (say 360) and a unity as being itself a member of a class.

The relation spoken of above in which two classes possessing the same number stand to one another does not involve counting. Think of the fingers on your hands. If to every finger of each hand belongs, by some process of correspondence, one and only one—remember the above meaning of this phrase—of the other, they are said to have "the same number." This is a definition of what "the same number" is to mean for us; the word "number" by itself is to have, as yet, no meaning for us; and, to avoid confusion, we had better replace the phrase "have the same number" by the words "are equivalent." Any other word would, of course, do, but this word happens to be fairly suggestive and customary. Now, if the variable u is any class, "the number of u" is defined as short for the phrase: "the class whose members are classes which are similar to u." Thus the number of u is an entity which is purely logical in its nature. Some people might urge that by "number" they mean something different from this, and that is quite possible. All that is maintained by those who agree to the process sketched above is: (1) Classes of the kind described are identical in all known arithmetical properties with the undefined things people call "integer numbers"; (2) It is futile to say: "These classes are not *numbers*," if it is not also said what *numbers* are—that is to say, if "the number of" is not defined in some more satisfactory way. There may be more satisfactory definitions, but this one is a perfectly sound foundation for all mathematics, including the theory not touched upon here of *ordinal numbers* (denoted by "first," "second," . . .) which apply to sets arranged in some order, known at present.

To illustrate (1), think of this. Acording to the above definition 2 is the general idea we call "couple." We say: "Mr. and Mrs. A. are a couple"; our definition would ask us to say in agreement with this: "The class consisting of Mr. and Mrs. A. is a member of the class 2." We define "2" as "the class of classes u such that, if x is a u, u lacking x is a 1"; the definition of "3" follows that of "2"; and so on. In the same way, we see that the class of fingers on your right hand and the class of fingers on your left hand are each of them members of the class 5. It follows that the classes of the fingers are equivalent in the above sense.

Out of the striving of human minds to reproduce conveniently and anticipate the results of experience of geometrical and natural events, mathematics has developed. Its development gave priceless hints to the development of logic, and then it appeared that there is no gap between the science of number and the science of the most general relations of objects of thought. As for geometry and mathematical physics, it becomes possible clearly to separate the logical parts from those parts which formulate the data of our experience.

We have seen that mathematics has often made great strides by sacrificing accuracy to analogy. Let us remember that, though mathematics and logic give the highest forms of certainty within the reach of us, the process of mathematical discovery, which is so often confused with what is discovered, has led through many doubtful analogies and errors arriving from the great help of symbolism in making the difficult easy. Fortunately symbolism can also be used for precise and subtle analysis, so that we can say that it can be made to show up the difficulties in what appears easy and even negligible—like $1 + 1 = 2$. This is what much modern fundamental work does.

CHAPTER VII

THE NATURE OF MATHEMATICS

IN the preceding chapters we have followed the development of certain branches of knowledge which are usually classed together under the name of "mathematical knowledge." These branches of knowledge were never clearly marked off from all other branches of knowledge: thus geometry was sometimes considered as a logical study and sometimes as a natural science—the study of the properties of the space we live in. Still less was there an absolutely clear idea of what it was that this knowledge was about. It had a name—"mathematics"—and few except "practical" men and some philosophers doubted that there was something about which things were known in that kind of knowledge called "mathematical." But what it was did not interest very many people, and there was and is a great tendency to think that the question as to what mathematics is could be answered if we only knew all the facts of the development of our mathematical knowledge. It seems to me that this opinion is, to a great extent, due to an ambiguity of language: one word—"mathematics"—is used both for our knowledge of a certain kind and the thing, if such a thing there be, about which this knowledge is. I have distinguished, and will now explicitly distinguish, between "Mathematics," a collection of truths of which we know something, and "mathematics," our knowledge of Mathematics. Thus, we may speak of "Euclid's mathematics" or "Newton's mathematics," and say truly that mathematics has developed and therefore had history; but Mathematics is eternal and unchanging, and therefore has no history—it does not belong, even in part, to Euclid or Newton or anybody else, but is something which is discovered, in the course of time, by human minds. An analogous distinction can be drawn between "Logic" and "logic." The small initial indicates that we are writing of a psychological process which may lead to Truth; the big initial indicates that we are writing of the entity—the part of Truth—to which this process leads us. The reason why mathematics is important is that Mathematics is not incomprehensible, though it is eternal and unchanging.

Grammatical usage makes us use a capital letter even for "mathematics" in the psychological sense when the word begins a sentence, but in this case I have guarded and will guard against ambiguity.

That particular function of history which I wish here to emphasise will now, I think, appear. In mathematics we gradually learn, by getting to know some thing about mathematics, to know that there is such a thing as Mathematics.

We have, then, glanced at the mathematics of primitive peoples, and have seen that at first discoveries were of isolated properties of abstract things like numbers or geometrical figures, and of abstract relations between concrete things like the relations between the weights and the arms of a lever in equilibrium. These properties were, at first, discovered and applied, of course, with the sole object of the satisfaction of bodily needs. With the ancient Greeks comes a change in point of view which perhaps seems to us, with our defective knowledge, as too abrupt. So far as we know, Greek geometry was, from its very beginning, deductive, general, and studied for its own interest and not for any applications to the concrete world it might have. In Egyptian geometry, if a result was stated as universally true, it was probably only held to be so as a result of induction —the conclusion from a great number of particular instances to a general proposition. Thus, if somebody sees a very large number of officials of a certain railway company, and notices that all of them wear red ties, he might conclude that *all* the officials of that company wear red ties. This might be probably true: it would not be certain: for *certainty* it would be necessary to know that there was some rule according to which all the officials were compelled to wear red ties. Of course, even then the conclusion would not be certain, since these sort of laws may be broken. Laws of *Logic*, however, cannot be broken. These laws are not, as they are sometimes said to be, laws of *thought*; for logic has nothing to do with the way people think, any more than poetry has to do with the food poets must eat to enable them to compose. Somebody might *think* that 2 and 2 make 5: we know, by a process which rests on the laws of Logic, that they make 4.

This is a more satisfactory case of induction: Fermat stated that no integral values of x, y, and z can be found such that $x^n + y^n = z^n$, if n be an integer greater than 2. This theorem has been proved to be true for $n = 3, 4, 5, 7$, and many other numbers, and there is no reason to doubt that it is true. But to this day no general proof of it has been given.[8] This, then, is an example of a mathematical proposition which has been reached and stated as probably true by induction.

Now, in Greek geometry, propositions were stated and proved by the

[8] This is an example of the "theory of numbers," the study of the properties of integers, to which the chief contributions, perhaps, have been made by Fermat and Gauss.

laws of Logic—helped, as we now know, by tacit appeals to the conclusions which common sense draws from the pictorial representation in the mind of geometrical figures—about *any* triangles, say, or *some* triangles, and thus not about one or two particular things, but about an *infinity* of them. Thus, consider any two triangles *ABC* and *DEF*. It helps the thinking of most of us to draw pictures of particular triangles, but our conclusions do not hold merely for these triangles. If the sides *BA* and *AC* are equal in length to the sides *ED* and *DF* respectively, and the angle at *A* is equal to the angle at *D*, then *BC* is equal to *EF*. This is proved rather imperfectly in the fourth proposition of the first Book of Euclid's *Elements*.

When we examine into and complete the reasonings of geometricians, we find that the conception of space vanishes, and that we are left with logic alone. Philosophers and mathematicians used to think—and some do now—that, in geometry, we had to do, not with the space of ordinary life in which our houses stand and our friends move about, and which certain quaint people say is "annihilated" by electric telegraphs or motor cars, but an abstract form of the same thing from which all that is personal or material has disappeared, and only things like *distance* and *order* and *position* have remained. Indeed, some have thought that position did not remain; that, in abstract space, a circle, for example, had no position of its own, but only with respect to other things. Obviously, we can only, in practice, give the position of a thing with respect to other things—"relatively" and not "absolutely." These "relativists" denied that position had any properties which could not be practically discovered. Relativism, in a thought-out form, seems quite tenable; in a crude form, it seems like excluding the number 2, as distinguished from classes of two things, from notice as a figment of the brain, because it is not visible or tangible like a poker or a bit of radium or a mutton-chop.

In fact, a perfected geometry reduces to a series of deductions holding not only for figures in space, but for any abstract things. Spatial figures give a striking illustration of some abstract things; and that is the secret of the interest which analytical geometry has. But it is into algebra that we must now look to discover the nature of Mathematics.

We have seen that Egyptian arithmetic was more general than Egyptian geometry: like algebra, by using letters to denote unknown numbers, it began to consider propositions about *any* numbers. In algebra and algebraical geometry this quickly grew, and then it became possible to treat branches of mathematics in a systematic way and make whole classes of problems subject to the uniform and almost mechanical working of one method. Here we must again recall the economical function of science.

At the same time as methods—algebra and analytical geometry and the infinitesimal calculus—grew up from the application of mathematics to natural science, grew up also the new conceptions which influenced the form which mathematics took in the seventeenth, eighteenth, and nineteenth centuries. The ideas of *variable* and *function* became more and more prominent. These ideas were brought in by the conception of motion, and, unaffected by the doubts of the few logicians in the ranks of the mathematicians, remained to fructify mathematics. When mathematicians woke up to the necessity of explaining mathematics logically and finding out what Mathematics is, they found that, in mathematics the striving for generality had led, from very early times, to the use of a method of deduction used but not recognised and distinguished from the method usually used by the Aristotelians. I will try to indicate the nature of these methods, and it will be seen how the ideas of variable and function, in a form which does not depend on that particular kind of variability known as motion, come in.

A *proposition* in logic is the kind of thing which is denoted by such a phrase as: "Socrates was a mortal and the husband of a scold." If—and this is the characteristic of modern logic—we notice that the notions of variable and function (correspondence, relation) which appeared first in a special form in mathematics, are fundamental in all the things which are the objects of our thought, we are led to replace the particular conceptions in a proposition by variables, and thus see more clearly the structure of the proposition. Thus: "x is a y and has the relation R to z, a member of the class u" gives the general form of a multitude of propositions, of which the above is a particular case; the above proposition may be true, but it is not a judgment of logic, but of history or experience. The proposition is false if "Kant" or "Westminster Abbey" is substituted for "Socrates": it is neither if "x," a sign for a variable, is, and then becomes what we call a "propositional function" of x and denote by "ϕx" or "ψx." If more variables are involved, we have the notation "$\phi(x,y)$," and so on.

Relations between propositional functions may be true or false. Thus x is a member of the class a, and a is contained in the class b, together imply that x is a b, is true. Here the *implication* is true, and we do not say that the *functions* are. The kind of implication we use in mathematics is of the form: "If ϕx is true, then ψx is true"; that is, any particular value of x which makes ϕx true also makes ψx true.

From the perception that, when the notions of variable and function are introduced into logic, as their fundamental character necessitates, all mathematical methods and all mathematical conceptions can be defined in purely logical terms, leads us to see that Mathematics is only a part of Logic and is the class of all propositions of the form: $\phi(x,y,z, \ldots)$ implies, for all values of the variables, $\psi(x,y,z, \ldots)$. The structure of the propositional functions involves only such ideas as are fundamental in logic, like implication, class, relation, the relation of a term to a class of which it is a member, and so on. And, of course, Mathematics depends on the notion of Truth.

When we say that "$1 + 1 = 2$," we seem to be making a mathematical statement which does not come under the above definition. But the statement is rather mistakenly written: there is, of course, only *one* whole class of unit classes, and the notation "$1 + 1$" makes it look as if there were two. Remembering that 1 is a class of certain classes, what the above proposition means is: If x and y are members of 1, and x differs from y, then x and y together make up a member of 2.

At last, then, we arrive at seeing that the nature of Mathematics is independent of us personally and of the world outside, and we can feel that our own discoveries and views do not affect the Truth itself, but only the extent to which we or others see it. Some of us discover things in science, but we do not really create anything in science any more than Columbus created America. Common sense certainly leads us astray when we try to use it for the purposes for which it is not particularly adapted, just as we may cut ourselves and not our beards if we try to shave with a carving knife; but it has the merit of finding no difficulty in agreeing with those philosophers who have succeeded in satisfying themselves of the truth and position of Mathematics. Some philosophers have reached the startling conclusion that Truth is made by men, and that Mathematics is created by mathematicians, and that Columbus created America; but common sense, it is refreshing to think, is at any rate above being flattered by philosophical persuasion that it really occupies a place sometimes reserved for an even more sacred Being.